SPACECAM

W9-DGO-548

David and Charles

SPACECAM

IN CO-OPERATION WITH
NASA

TERRY HOPE

PHOTOGRAPHING THE FINAL FRONTIER – FROM APOLLO TO HUBBLE

A DAVID & CHARLES BOOK
Copyright © David & Charles Limited 2005, 2007

David & Charles is an F+W Publications Inc. company
4700 East Galbraith Road
Cincinnati, OH 45236

First published in 2005
Reprinted 2005
First paperback edition 2007

Text copyright © Terry Hope 2005, 2007

ISBN-13: 978-0-7153-2164-5 hardback
ISBN-10: 0-7153-2164-1 hardback

ISBN-13: 978-0-7153-2739-5 paperback
ISBN-10: 0-7153-2739-9 paperback

Printed in Singapore by KHL Printing
for David & Charles
Brunel House, Newton Abbot, Devon

Commissioning Editor: Neil Baber
Editor: Jennifer Fox-Proverbs
Art Editor: Mike Moule
Designer: Sarah Clark
Production Controller: Kelly Smith

Visit our website at www.davidandcharles.co.uk

David & Charles books are available from all good bookshops;
alternatively you can contact our Orderline on 0870 9908222
or write to us at FREEPOST EX2 110, D&C Direct, Newton Abbot,
TQ12 4ZZ (no stamp required UK only); US customers call
800-289-0963 and Canadian customers call 800-840-5220.

TERRY HOPE is an award-winning journalist and author. He began his career writing for *Camera Weekly* and *Amateur Photographer*, and has since written for numerous photographic magazines. He is also a regular contributor to *The Times*, the *Sunday Telegraph* and *The Guardian* on a variety of subjects. He is the author of a number of books on photography including *Earthcam* (David & Charles) and is currently editor of *Professional Photographer* magazine.

CONTENTS

Man has always been fascinated by what lies beyond the boundaries of this planet, and space exploration, albeit at a level that was dictated by the capabilities of the equipment that was available, has been going on for hundreds of years. The invention of photography in the early 19th century was a crucial breakthrough for astronomers, who saw it as being a natural partner to their studies, allowing them to record detailed images from life that could subsequently be used for research and to back up findings that they had made.

The development of space travel since the war has opened up an exciting new range of opportunities, with imaging technology keeping pace with the breakthroughs that have been taking place in spacecraft and telescope design. The result has been the production of a stream of pictures that are extraordinary in their clarity and detail and, thanks to the benefits of the internet, the information is there as a priceless resource, available for schools and interested individuals to tap into, greatly increasing our knowledge and understanding of the universe and emphasizing our tiny role within it.

The invention of photography brought with it the promise of factual, fully detailed records of planetary observations that would carry far more weight than the hand-drawn references that had been used up to that point. Louis Daguerre, who was one of the pioneers of photography, perfecting the Daguerreotype in 1837, is also believed to have made the first photographic images of the moon, although these were said by contemporaries to be 'faint and lacking in detail.' Others who followed him were more successful. A Lunar Daguerreotype made by Whipple and Bond early in 1851 at the Harvard Observatory, was exhibited at the Great International Exhibition of the same year, and became a celebrated object in its own right, being praised for its clarity and detail, which was far superior to anything that had gone before. In the same year Whipple and Bond also managed to photograph two of Jupiter's bands, while photographer and astronomer Warren De la Rue produced images of Mars in 1856 and, two years later, he announced that he had pictures of Jupiter, Saturn and the double star Alpha Geminorum as well.

Thus the close relationship between photography and space was forged early on, and it has grown and matured ever since. In the exciting years since the Second World War, when man first started to think in earnest of leaving the planet and exploring space, photography has been the prime means by which these adventures have been recorded. Cameras captured the first views of Earth from manned orbit, the first daring walks in space and those tentative first steps on the surface of the Moon, and through dramatic photographs the world shared the experience.

Photographing man in space

Photography was not given a high priority in the first two sub-orbital flights of American astronauts because it was essential that they concentrate on operating the spacecraft. It was only with the Mercury-Atlas 6 mission in February 1962, when John H Glenn Jnr made the first orbital flight, that a hand-held camera was provided, this being the commercially available 35mm Ansco Autoset.

From the outset it was realized that photography had a vital role to play in recording missions, but it was equally obvious that astronauts would face major restrictions if they tried to take pictures with conventional gear. Heavy gloves made it virtually impossible to operate the controls of the camera, while astronauts in the early spacecraft were required to take pictures without removing their helmets, and so normal composition through the viewfinder was also a challenge. These problems were overcome by making the camera fully automatic to do away with the need to adjust shutter speed and aperture, and fitting it with a pistol grip and an oversized film advance.

A new type of viewfinder was also designed, which allowed the eye to be further from the camera, while still giving a good indication of the view that was being acquired.

For the Mercury-Atlas 7 Mission in 1962 the decision was made to continue with 35mm cameras, although this time the model taken into space was a Robot Recorder with well developed automatic film advance features. This same camera was also taken on the Mercury-Atlas 9 flight in 1963, being modified for dim-light photographic experiments by the provision of a fixed lens with a large f/0.95 aperture. Exposures were timed manually, and three small supports, or 'feet,' were provided to help the operator to position the camera against the window of the spacecraft.

One of the most significant developments in early space photography was actually made on the preceding Mercury-Atlas 8 mission, following the realization that film format could play a crucial role in improving the quality of the results being achieved. In simplistic terms, a 6x6cm negative could carry a great deal more information than one 35mm in size, and if the drawbacks of using a medium format system could be overcome – i.e., the fact that the cameras were more cumbersome and usually only capable of providing around 12 images before the film had to be changed – then the data recorded would be significantly enhanced.

Thus began the long association between NASA and the Swedish-designed Hasselblad camera. A serious professional tool, the Hasselblad offered not just a larger format but all the advantages of interchangeable lenses and even interchangeable backs, allowing astronauts to switch between black and white and colour film easily, and to utilize different film ISO speeds. The camera was a single lens reflex model, but by stripping out the reflex mirror arrangement, which was difficult to utilize in space in any case, and providing a straight eye-level viewfinder, ease of use, even for a fully kitted astronaut, was guaranteed.

There was still the problem of the short number of exposures to solve, and it was realized that the backing paper on the film could be dispensed with in the name of efficiency. Thus the camera was adapted to take a newly developed 70mm film back, allowing it to accommodate 18 feet of 70mm film, enough for 70 exposures to be taken. Better still, by applying the conventional film emulsion to a specially designed ultra-thin polyester base the same 70mm back, with some modifications, could hold 38 to 42 feet of film, which was enough for approximately 200 exposures. The development immediately took the quality of NASA'a photography on to a new level, and Hasselblad cameras, albeit in a much modified form, accompanied astronauts on missions throughout the heady years of the sixties and beyond. A modified 500EL model even made it to the surface of the Moon, being used by Neil Armstrong to record the first-ever visit by man.

Mission photography today

Photography is a key part of every NASA mission, and astronauts are trained to provide images that will provide vital information to scientists and others back on Earth. Joseph R 'Joe' Tanner has been with NASA since being selected as an astronaut candidate in March 1992, and he has extensive space flight experience, having completed three shuttle missions. He also performed two space walks as a member of the STS-82 crew whose task it was to service the Hubble Space Telescope in February 1997, and was involved in the assembly of the International Space Station on mission STS-97, completing a further three space walks during the ten days that he was in space on that occasion.

'The photographic equipment we use in space these days has not been modified in any significant way,' he says. 'The only ones that are a little different are the ones that we take outside the spacecraft: these are specified EVA (Extra Vehicular Activity) lenses, and we change the lubrication on them to one that's vacuum capable, so that they will still autofocus, and the moving parts will continue to function.

'Photography is a very important part of what I do. On my first mission on the Space Shuttle Atlantis in 1994 I was on a flight that was due to study the Earth's

atmosphere and solar effects during the Sun's 11-year cycle, and we took about 10,000 frames. Most of my experience has been with the 70mm Hasselblads, large format Linhof cameras and 35mm equipment, with digital cameras now coming into the frame.

'The 35mm cameras I have used have all been Nikons, latterly Nikon F5s and before that F4s, and, for some of the EVA work, F3s. The F3s were used in strictly manual mode, and we'd fix the focus and aperture by taping the lens so that it couldn't move, taking care to keep a certain distance from our subject so that we retained focus. The lenses for 35mm work consist of a straight 50mm that we use for most of the cabin shots, and then we have a couple of short to mid-range zooms and a fixed 180mm. We also have a fixed 400mm lens and we can put a doubler on this to turn it into an 800mm, and that's probably our prime lens for Earth observations. With the Hasselblad camera we have 50mm, 110mm and 250mm lenses.

'The digital cameras we are using are Kodak 760 models that are based on the Nikon F5, and these have now been cleared for EVA work. These are big 6 megapixel cameras that are not state of the art, but it's the one that we have around and it's certified to go into space. The big advantage of using digital for us is that pictures can be sent back to NASA while we are still in space. Film, once it's been exposed, has to be brought back to Earth for processing really quickly, otherwise it will become fogged thanks to the number of cosmic hits that it takes. This is particularly the case with the faster films, such as ISO 400 and 800.

'If we are taking pictures of the Earth in daylight then we could be shooting at shutter speeds of 1/250sec to 1/500sec, and you should be pretty safe hand holding the camera under these circumstances. If you're taking a low light shot, however, things are more difficult, and we can use a mechanical arm that is very stable, although this can be a little hard to set up sometimes. We can also use something called a lockline, which is a piece of flexible plastic that wouldn't hold the weight of the camera on Earth, but in space it holds it just fine. There are other considerations to bear in mind when you are photographing the Earth from a spacecraft in orbit. You have to be very cautious about stray light in the cockpit if you're taking a dark shot, because this will reflect off the glass in the window and ruin the shot. You also quickly realize that your target is moving: you are travelling at five miles a second, and if you are using a slower shutter speed you need to perfect the art of panning the camera to keep track with the movement of the Earth and to maintain focus.

'All of us have received a lot of photographic training, and we have training cameras that are identical to the cameras that we take into space. When you are first or second crew you get a locker that has every type of camera that we fly, and we can take one home at any time and practise with it. You tend to get a lot of good family shots while you're becoming acquainted with the cameras!

'There are a number of reasons why we need to take pictures in space. On the International Space Station, there are times when we need to document things on the inside of the vehicle to explain to the ground what a particular problem might be, and this is probably the most critical use of the camera. There is also a public relations aspect, in that photography allows us to document life on board, and allows people to see what the crew is doing, which keeps them interested.

'In terms of pictures of Earth, NASA has been doing this now for over 40 years, and it's very interesting to compare locations and the way they have changed in this time. We have a very professional Earth Observation Department, and it's their full-time job to look at this kind of photography and to analyse it. In locations such as Africa you can track the deforestation and the encroachment of desert regions, and see how crops might be bringing subterranean water to the surface, thereby decreasing the water table in these areas. You can look at areas that you couldn't travel to conventionally very easily, and research of this kind is a vital tool for the people who are monitoring the health of our planet.'

Photographing deeper space

Photography has also allowed us to see views from space that no human has seen first hand. The Viking 1 and Viking 2 missions for example, both launched in 1975, provided startling views of the surface of Mars that were better than had ever been seen before, transmitting back 1400 images between them. The Voyager 1 and Voyager 2 spacecraft, launched in 1977, delivered still more spectacular results, both visiting Jupiter and Saturn, with Voyager 2 going on to encounter Uranus and Neptune as well. The pictorial results that the craft supplied were exceptional, and images were received back until 1990.

More recently there have been remarkable images sent back by the two Mars Exploration Rover Missions, which were launched in 2003, both landing successfully on the surface of Mars in early 2004. Part of a long-term aim to provide robotic exploration of the red planet and to determine whether there was once water activity on the planet's surface, the visual information sent back over a two year period by the two rovers has helped immeasurably in the preparation for an eventual manned landing. Another spectacular recent success has been the Cassini-Huygens mission to Saturn and its moon Titan. Saturn's beautiful rings, extending for thousands of miles from the planet, are the most extensive and complex in our solar system, and they are made up of billions of particles of ice and rock, which range in size from grains of sugar to

houses. Saturn's 34 known moons are equally mysterious, especially Titan. Bigger than the planets Mercury and Pluto, Titan is of particular interest to scientists because it is one of the few moons in our solar system with its own atmosphere. The moon is cloaked in a thick, smog-like haze that scientists believe may be very similar to Earth's before life began more than 3.8 billion years ago.

Four NASA spacecraft have been sent to explore Saturn. Pioneer 11 was first to fly past the planet in 1979, Voyager 1 flew past a year later, followed by its twin, Voyager 2, in 1981. The Cassini spacecraft, however, was the first to explore the Saturn system of rings and moons from orbit, and it began this phase of its mission on June 30, 2004 and immediately began sending back intriguing images and data. The European Space Agency's Huygens Probe, attached to Cassini, had its own agenda, plunging into Titan's thick atmosphere in January 2005, and sending back the first-ever images seen of this moon's surface.

The Hubble Space Telescope has been delivering pictures that peer still further into the depths of the universe since it was deployed by the space shuttle in 1990. Packed with sensitive and high-powered instruments and free of the restraints of Earth's atmosphere, Hubble has lived up to all the expectations of the scientists who spent 15 years on its development. The pictures it has returned during its lifetime are extraordinarily beautiful and have revolutionized our understanding of deep space.

Other satellites orbiting the Earth have been pointed towards us, and have revealed aspects of our planet that could never be observed in any other way. Weather conditions can be monitored on a continual basis, extreme events such as hurricanes can be tracked, and even volcanoes can be recorded during times of eruption to enable their progress to be documented and the wider picture to be revealed. The all-seeing eye in space can observe the break up of ice floes and report on the effects of global warming, and even give us sobering details of major catastrophes, such as the devastating tsunami at the end of 2004.

Small wonder that, with so many exciting developments occurring, there has never been greater interest in space photography of all kinds. Spacecam has pulled together some of the most striking images from the past 40 years, and presents them here as a record of NASA's achievements in space in that time. It has been an extraordinary period of discovery, with the promise of much more to come, and the pictures here, while giving us so much vital scientific information, are also united by their great beauty. To see them is to acknowledge our own vulnerability, and the insignificance of Earth within the vast expanse of the solar system.

TERRY HOPE

PART ONE: LOOKING OUT

The night sky has drawn the gaze of man since the earliest times and observations carefully made and recorded. Early telescopes allowed even more of the universe to be visually explored but the invention of photography in the early 19th century was a crucial advance, as astronomers quickly realized that this was the means by which they could accurately record and share some of the marvels they were seeing. By the start of the 20th century highly detailed images of the Moon and the closer planets were widely available.

The dramatic advances that have really pushed back the boundaries of our knowledge, however, have taken place over the past half century, with the successful conquest of the Moon, and the regular, almost routine, travel out of Earth's atmosphere by the US and Russian space programmes and the associated development of the International Space Station. Further to these, there has been the launch of several notable missions into the deeper reaches of the galaxy that have sent back photographs of such quality that even those of just a generation ago would have struggled to believe them.

Those who lived through the time of the first expeditions to the Moon will never forget the wave of excitement that accompanied this great adventure, and the amazement at the pictures that those pioneering explorers brought back with them. In more recent times the focus has been further afield, with vast numbers of breath-taking images being sent back from the surface of Mars after the Mars Exploration Rover Missions landed in 2004, while Saturn and its amazing rings and 34 moons – including Titan, which is of particular interest to scientists because it is one of the few moons in our solar system with its own atmosphere – have been revealed in all their glory, giving up their secrets and expanding our knowledge immeasurably.

Fresh discoveries are being made at regular intervals, and space is, as it should be, regular headline news, with dramatic images making the front pages, and encouraging the public to take a real interest in the vast world beyond our home planet. There is, however, a limit to how far man can travel away from the Earth at the moment, and even the most brilliantly designed of spacecraft, such as Cassini, can't hope to reach even the closest star. This is where the Hubble Space Telescope has really come into its own, being devised as the realization of a dream by Earth-bound astronomers who had longed since 1946 to unleash the potential of a space telescope that could see further into the distant reaches of the universe than ever before.

With early teething problems overcome, Hubble has been a spectacular success, producing a succession of fantastic images that have fired the imagination of a new generation, giving us the ability to see, in extraordinary detail, celestial bodies and galaxies that were previously hidden from our view. Some of the images have pulled detail in from so far away that what we are seeing dates back close to the formation of the universe itself: it's a rare privilege for the modern human being to witness such material, and it is expanding our knowledge and giving us a taste to find out even more.

Hubble follows the traditional style of any telescope, being basically a long tube that is open at one end, and the distribution of the light inside the device is controlled by mirrors that are made of glass and coated with layers of pure aluminium (three-millionths of an inch thick) and magnesium fluoride (one millionth of an inch thick), which allows them to reflect visible, infrared and ultraviolet light. Light enters the telescope through the opening and bounces off the primary mirror to a secondary mirror. This then reflects the light through a hole in the centre of the primary mirror to a focal point behind it, where smaller, half-reflective, half-transparent mirrors distribute the light to the various scientific instruments.

There are five of these, each one of which has a specialized function to perform, and they have been regularly updated throughout Hubble's lifetime. The instruments are capable of discerning different wavelengths of light, such as visible, ultraviolet and infrared, and monitoring each of these allows different properties of objects in the universe to be studied. Conventional film has never been used on board Hubble, and instead images are captured digitally using charge-coupled devices (CCDs), which are stored in on-board computers and subsequently relayed to Earth.

The current generation is privileged indeed to be alive at a time when so many exciting discoveries are being made, with the certainty is that the next fifty years will be even more awe-inspiring. It's quite an adventure that is being promised, and the imagination can't help but be stretched at the thought of what is yet to come.

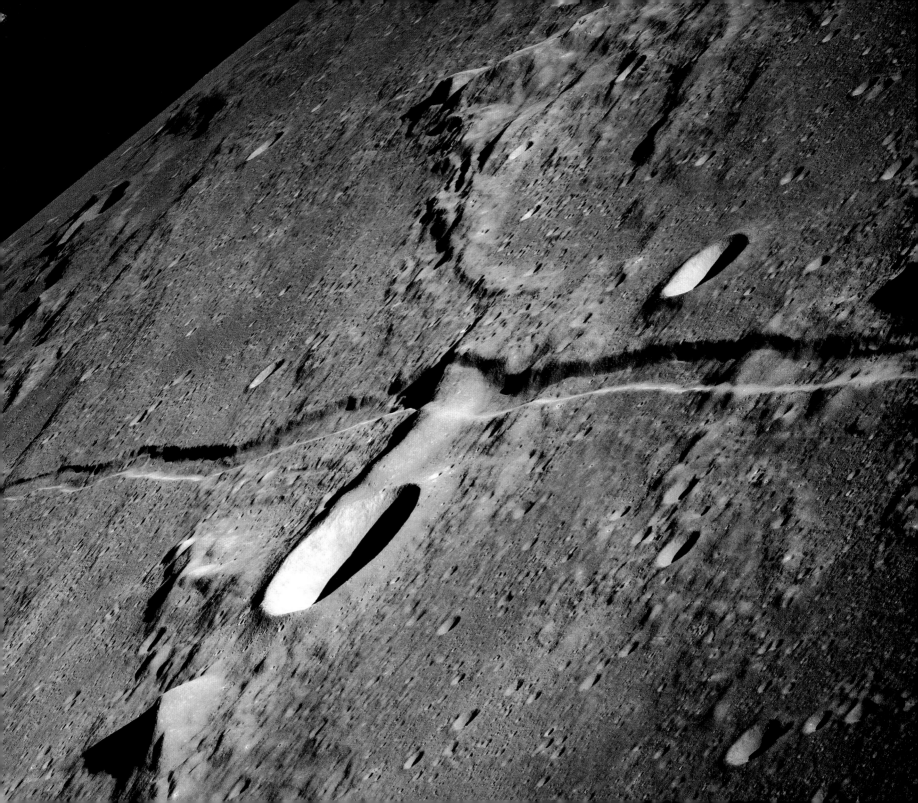

MAN IN SPACE:
THE MOON

PREVIOUS PAGE
This dramatic oblique view of the Moon's surface was photographed by the Apollo 10 astronauts in May 1969, using a handheld 70mm camera and a 250mm lens.

This high forward oblique view of the Rima Ariadaeus area of the Moon was photographed by an Apollo 10 astronaut, who aimed a handheld 70mm camera at the surface during lunar orbit.

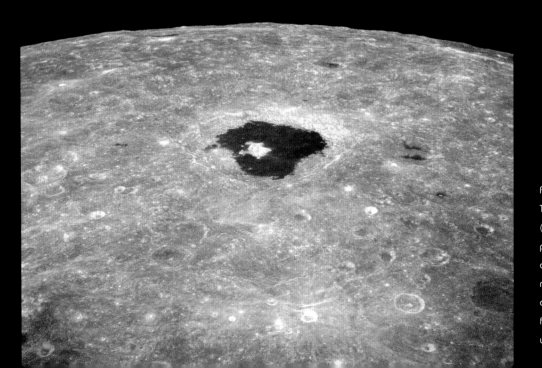

A view of the Moon's large crater Tsiolkovsky, some 241.5 kilometres (150 miles) in diameter, as photographed by the astronauts during the Apollo 8 lunar orbit mission. The site was first identified and named by the Russians from photographs taken by their unmanned Luna III spacecraft.

The Apollo 16 Command and Service Module Casper approaches the Lunar Module as the two craft prepared to make their final rendezvous of the mission on April 23, 1972. Astronauts John W Young and Charles M Duke Jr aboard the Lunar Module were returning to Casper in lunar orbit after three successful days on the surface of the Moon.

PART ONE: LOOKING OUT

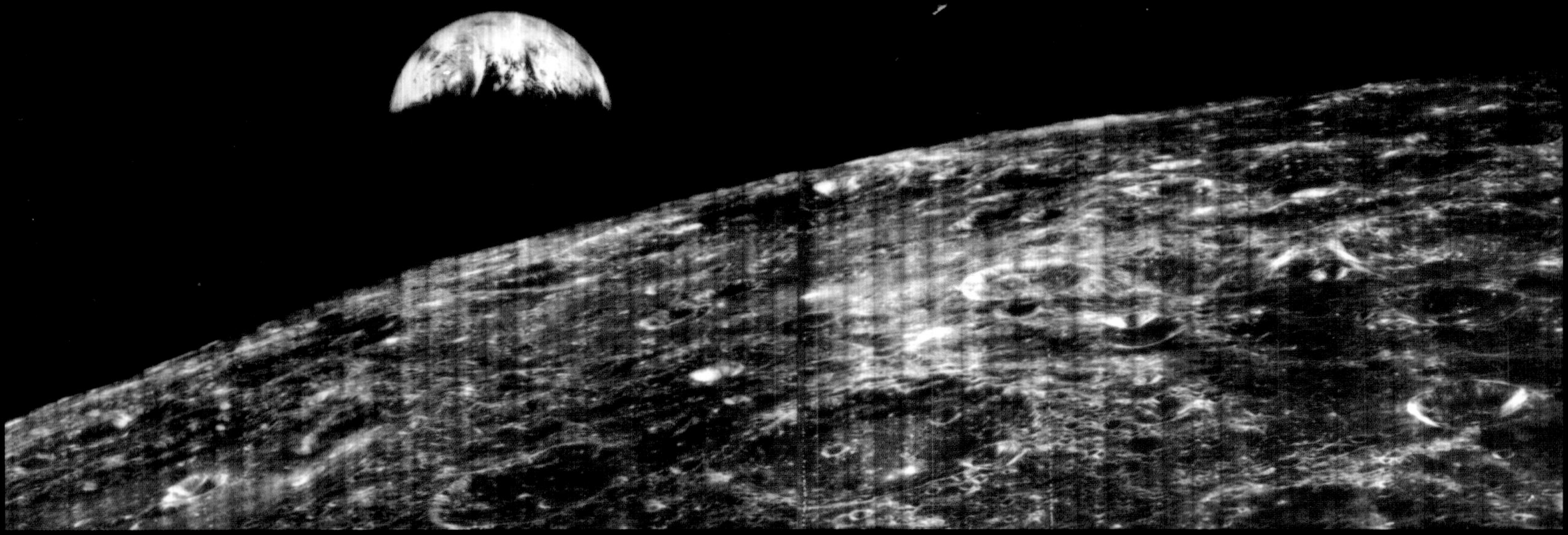

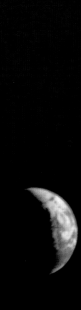

This picture of the Earth and Moon in a single frame, the

Edgar D Mitchell, Apollo 14's lunar module pilot, photographed this

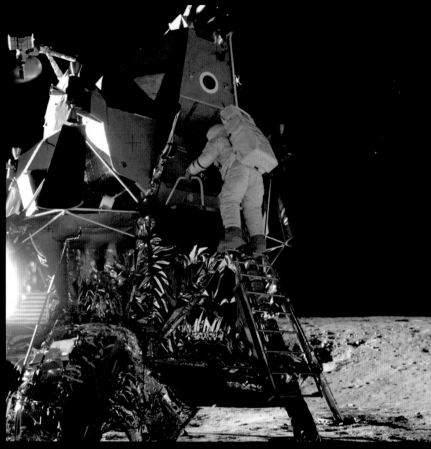

Edwin 'Buzz' Aldrin Jr, the pilot of the Lunar Module, is seen during the historic visit that Apollo 11 made to the Moon in 1969, having just deployed the Early Apollo Scientific Experiments Package. In the foreground is the Passive Seismic Experiment Package, and beyond it is the Laser Ranging Retro-Reflector. In the left background the black and white lunar surface television camera can be seen, while behind is Eagle, the Lunar Module that brought these first explorers to the Moon. Neil Armstrong, the commander of this mission, took the photograph using a 70mm lunar surface camera.

Alan L Bean, Lunar Module pilot for the Apollo 12 mission, starts down the ladder of the Lunar Module Intrepid to join mission commander Charles Conrad Jr on the lunar surface. November 1969.

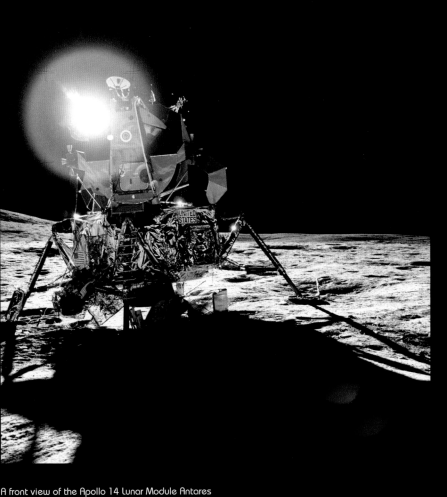

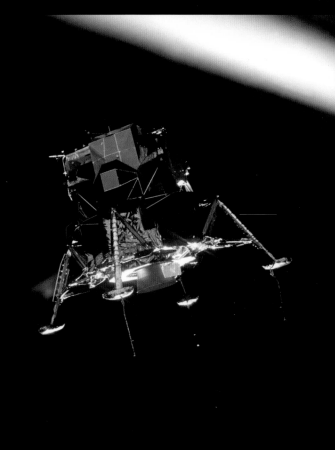

A front view of the Apollo 14 Lunar Module Antares which was reflecting a circular flare caused by the brilliant sun. This unusual ball of light was said by the astronauts to have a jewel-like appearance. January 1971.

The Apollo 11 Lunar Module Eagle seen in a landing configuration, was photographed in lunar orbit from the Command and Service Module Columbia, July 1969. The long rod-like protrusions that can be seen under the landing pods are lunar surface sensing probes. Upon contact with the lunar surface, the probes send a signal

Apollo 12 Lunar Module pilot Alan L Bean pauses near a tool carrier during one of the mission's Moon walks. Commander Charles Conrad Jr, who took the picture in November 1969, can be seen reflected in Bean's helmet visor.

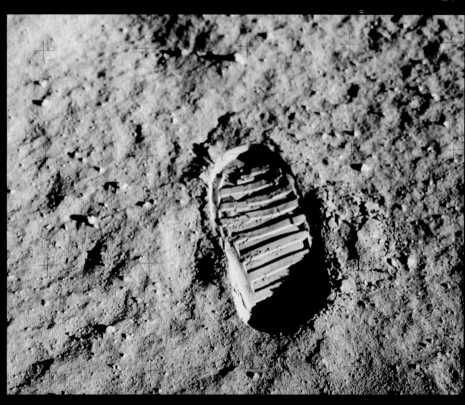

Alan L Bean, Lunar Module pilot for the Apollo 12 lunar landing mission, holds a Special Environmental Sample Container filled with lunar soil collected during a Moon walk undertaken with the mission's commander, Charles Conrad Jr. Conrad, who took this picture in the Ocean of Storms, can be seen reflected in Bean's helmet visor. November 1969.

A record of one of the first steps to be taken by humans on the Moon, this is an image of Buzz Aldrin's bootprint from the Apollo 11 mission. Neil Armstrong and Buzz Aldrin made their historic first walk on the Moon on July 20, 1969.

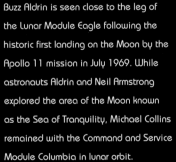

Buzz Aldrin is seen close to the leg of the Lunar Module Eagle following the historic first landing on the Moon by the Apollo 11 mission in July 1969. While astronauts Aldrin and Neil Armstrong explored the area of the Moon known as the Sea of Tranquility, Michael Collins remained with the Command and Service Module Columbia in lunar orbit.

Buzz Aldrin poses for a photograph beside the deployed United States flag during one of the first Moon walks. The Lunar Module is on the left, and the footprints of the astronauts are clearly visible in the soil of the Moon. July 1969.

This classic view of the rising Earth, which has become one of the signature images from space, greeted the Apollo 8 astronauts as they came from behind the Moon after the lunar orbit insertion burn. December 1968.

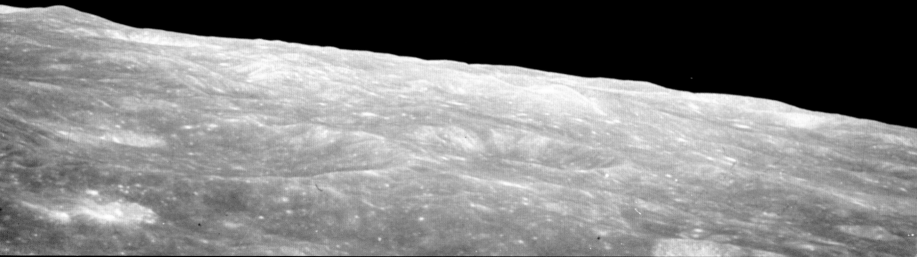

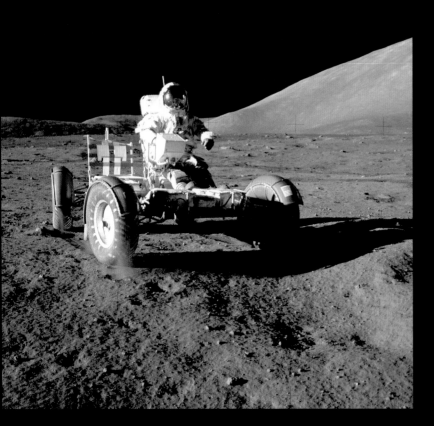

Eugene A Cernan, the commander of Apollo 17, makes a short checkout of the Lunar Roving Vehicle during the early part of the mission's first Moon walk at the Taurus-Littrow landing site in December 1972. The mountain in the right background is the east end of the South Massif.

Eugene A Cernan, commander of Apollo 17 salutes the flag on the Moon's surface during NASA's final lunar landing mission. The Lunar Module Challenger is in the left background behind the flag and the Lunar Roving Vehicle is also in the background behind him.

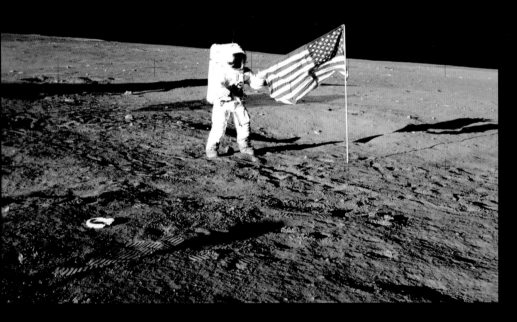

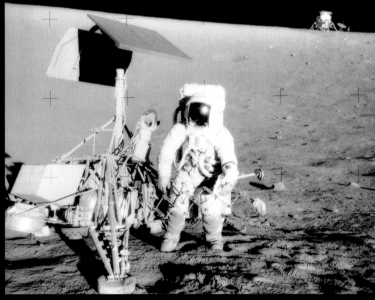

Apollo 12 astronaut Charles 'Pete' Conrad unfurls the United States flag on the lunar surface during the mission's first Moon walk in the Ocean of Storms on November 19, 1969. Several footprints made by the crew are visible in the photograph.

Charles Conrad Jr, commander of Apollo 12, examines the unmanned Surveyor III spacecraft during the mission's second Moon walk, with the Lunar Module Intrepid in the right background. The Intrepid landed on the Moon's Ocean of Storms only 183 metres (600 feet) from Surveyor III, which had soft landed on the Moon during April 1967. Its television camera and several other components were salvaged and brought back to earth for scientific analysis.

Charles M Duke Jr, the pilot of Apollo 16's Lunar Module, collects lunar samples at Station No 1 during the mission's first Moon walk at the Descartes landing site, April 1972. Duke is standing at the rim of Plum Crater, some 40 metres (131 feet) in diameter and 10 metres (33 feet) deep, while the parked Lunar Roving Vehicle can be seen in the left background.

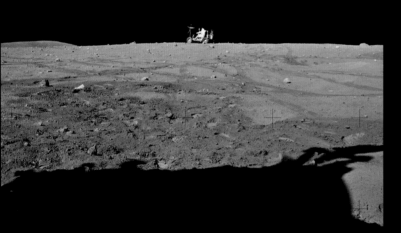

John W Young, the commander of Apollo 16, drives the Lunar Roving Vehicle to its final parking place near the end of the third expedition on to the Moon's surface at the Descartes landing site. The flank of Stone Mountain can be seen on the horizon at left, while the shadow of the Lunar Module Orion is visible in the foreground. April 1972.

Charles M Duke Jr, the pilot of the Lunar Module Orion on the Apollo 16 mission, stands near the Lunar Roving Vehicle at Station No 4, near Stone Mountain. The gnomon, a device used as a photographic reference to help determine the angle of the Sun, the scale within the scene and the lunar colour, is deployed in the centre foreground.

THE MOON

James B Irwin, the Lunar Module pilot for the Apollo 15 mission, works at the Lunar Roving Vehicle during its first expedition on the Moon's surface. The shadow of the Lunar Module Falcon is in the foreground, and the view was taken looking northeast, with Mount Hadley in the background.

Apollo 15's Lunar Module pilot James B Irwin loads up the Lunar Roving Vehicle with tools and equipment in preparation for the mission's first expedition on the lunar surface at the Hadley-Apennine landing site, July 1971. The Lunar Module Falcon can be seen to the left.

James B Irwin, the Lunar Module pilot for the Apollo 15 mission, uses a scoop to make a trench in the lunar soil.

The ascent of the Lunar Module Spider from the surface of the Moon was photographed from the Command and Service Module on the fifth day of the Apollo 9 earth-orbital mission, March 1969. The Lunar Module's descent stage had already been jettisoned.

The Lunar Roving Vehicle is
photographed alone against the
lunar background at the Hadley-
Apennine landing site of the Apollo 15

An extraordinary lunar panorama at
Station No 4 (Shorty Crater), showing
Geologist-Astronaut Harrison H Schmitt
working at the Lunar Roving Vehicle

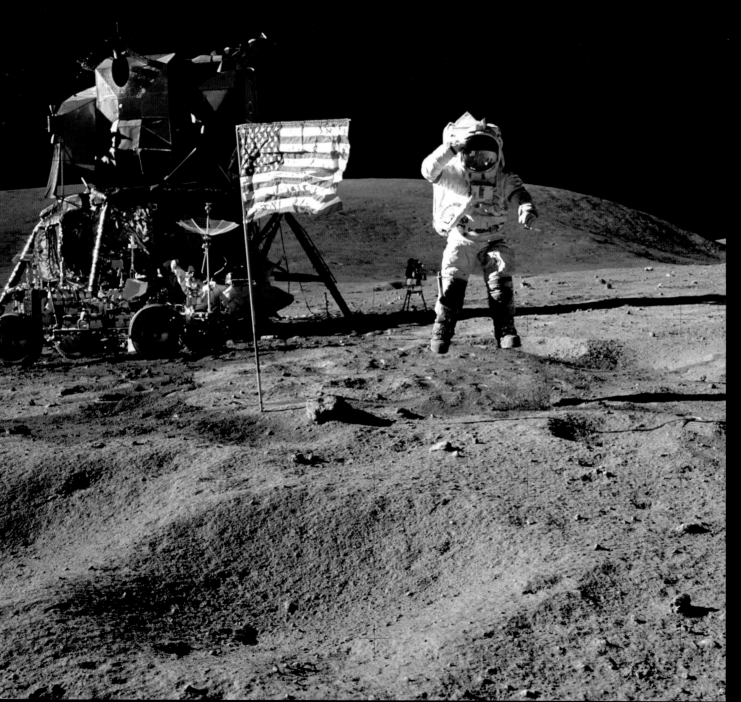

jumps up from the lunar surface as he salutes the United States Flag at the Descartes landing site during the mission's first Moon walk, April 1972. The Lunar Module Orion is on the left, while the Lunar Roving Vehicle is parked beside. The object behind Young in the shade of the Lunar Module is the Far Ultraviolet Camera Spectrograph.

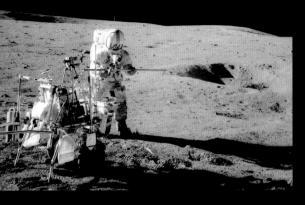

Apollo 14 commander Alan Shepard stands by the Modular Equipment Transporter, January 1971. The MET, which the astronauts nicknamed 'the rickshaw,' was a cart for carrying tools, cameras and sample cases on the lunar surface. Shepard can be identified by the vertical stripe on his helmet: after Apollo 13 (April 1970) the commander's spacesuit had red stripes on the helmet, arms, and one leg, to identify him in photographs.

John W Young, commander of the Apollo 16 mission, replaces tools in the hand tool carrier at the aft end of the Lunar Roving Vehicle during the mission's second expedition undertaken at the Descartes landing site. April 1972.

Astronaut Charles M Duke Jr, Lunar Module pilot during the Apollo 16 lunar landing mission, works at the Lunar Roving Vehicle, while all around him are scattered small rocks and boulders.

Harrison H Schmitt is seen next to a huge, split boulder at Station No 6 on the sloping base of North Massif during the third Apollo 17 Moon walk at the Taurus-Littrow landing site, December 1972.

Harrison Schmitt, the geologist and Apollo 17 Lunar Module pilot, uses an adjustable sampling scoop to retrieve lunar samples during the second lunar drive of the mission. The task has led to Schmitt's once immaculate spacesuit being covered with dirt. In the foreground a gnomon can be seen: some 46 centimetres (18 inches) long, this is used to provide an accurate vertical reference and calibrated length to allow the size and position of objects in near-field photographs to be established.

Astronaut Charles M Duke Jr took this picture of Apollo 16's Lunar Module Orion, which he had piloted to the surface of the Moon, from the moving Lunar Roving Vehicle. He was returning with Commander John W Young from the mission's third venture away from their craft, and his image captures the colour television camera mounted on the LRV in the foreground, along with a portion of the vehicle's high-gain antenna. April 1962

Apollo 15 commander David R Scott, with tongs and gnomon in hand, studies a boulder on the slope of Hadley Delta during the lunar rover expedition undertaken by the mission. The Lunar Roving Vehicle or Rover is in the right foreground. July 1971.

Commander David R Scott gives a military salute while standing beside the freshly-deployed American flag at Apollo 15's Hadley-Apennine landing site. The Lunar Module Falcon is partially visible on the right, while Hadley Delta in the background rises approximately 4,000 metres (about 13,124 feet) above the plain.

A view of the area at Station No 4 (Shorty Crater) showing the highly-publicized orange soil which the Apollo 17 crewmen found on the Moon during the second Apollo 17 Moon walk, December 1972.

During the historic first Moon walk of July 1969, Apollo 11's commander Neil A Armstrong is seen at the Modular Equipment Storage Assembly of the Lunar Module Eagle. The picture is a rarity, because most of the photographs from the Apollo 11 mission show Buzz Aldrin.

This oblique view featuring Crater 302 on the Moon's surface was photographed by the Apollo 10 astronauts in May 1969. The method was surprisingly simple: a member of the Apollo 10 crew aimed a handheld 70mm camera at the surface from lunar orbit for a series of pictures in this area.

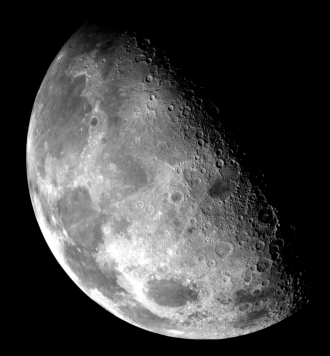

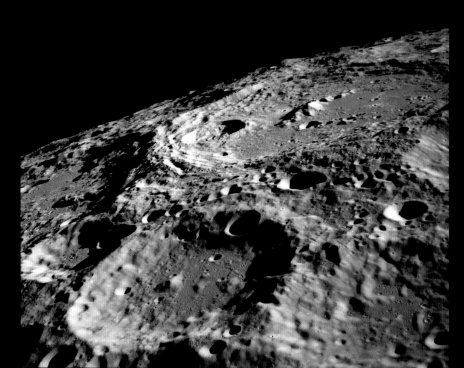

This view of the Moon's North Pole is a mosaic assembled from 18 images taken by the satellite Galileo's imaging system through a green filter as the spacecraft flew by on December 7, 1992. This area of the Moon is located just inside the shadow zone, about a third of the way from the top left of the illuminated region.

The Moon sets over the Earth's horizon, seen from the Space Shuttle Discovery during its STS-70 mission. July 1995.

MAN IN SPACE:
EARTH ORBIT

PREVIOUS PAGE
Apollo 9 Command and Service Module Gumdrop, and the Lunar Module
Spider, are shown docked together as Command Module pilot David R
Scott stands in the open hatch. At the time Apollo 9 was engaged on an
Earth orbital mission designed to test docking procedures between the
CSM and LM, as well as to test fly the Lunar Module in the relatively safe
confines of Earth's orbit.

Above left: The Space Shuttle Atlantis streaks skyward,
as sunlight pierces throuh the gap between the orbiter
and ET assembly. Atlantis lifted off on the 40-second
space shuttle flight at 11:02 am EDT on August 2, 1991
carrying a crew of five. A remote camera at the 84-metre
(275-foot) level of the Fixed Surface Structure took
this picture.

Scientist-astronaut Owen K Garriott, Skylab III science
pilot, attends to the Apollo Telescope Mount of the
Skylab space station, having just deployed the Skylab
Particle Collection S149 experiment to collect material
from interplanetary dust particles on prepared surfaces.
The picture was taken in 1973 by a fellow astronaut
using a hand- held 70mm Hasselblad camera.

Astronaut Jerry L Ross, anchored
to the foot restraint on the Remote
Manipulator System, approaches
the tower-like Assembly Concept
for Construction of Erectable Space
Structures device. The structure had
just been deployed by Ross and
astronaut Sherwood Spring as the
Atlantis flew over the white clouds
and blue ocean waters of the
Atlantic in 1985.

EARTH ORBIT

Mission Specialist Bruce McCandless II, is seen further away from the confines and safety of his ship than any previous astronaut has ever been. This space first was made possible by the Manned Manoeuvring Unit or MMU, a nitrogen jet propelled backpack. After a series of test manoeuvres inside and above Challenger's payload bay, McCandless went 'free-flying' to a distance of 320 feet (97.5 metres) away from the Orbiter.

Bruce McCandless II again, seen during his historic 1984 space walk. His MMU unit was controlled by joy sticks positioned at the end of the arm rests, and moving the joy sticks left or right or pulling them fired nitrogen jet thrusters to allow McCandless to move in any direction he chose. A still camera was mounted on the upper right portion of the MMU to allow him to record his surroundings.

On June 3, 1965 Edward H White II became the first American to step outside his spacecraft and let go, effectively setting himself adrift in the zero gravity of space. For 23 minutes he floated and manoeuvred himself around the Gemini spacecraft, being attached to a 25-ft umbilical line and a 23-ft tether line, both wrapped in gold tape to form one cord.

An ultimately unsuccessful individual attempt in 1992 by EVA mission specialist Pierre Thuot to capture Intelsat VI, as seen from the space shuttle Endeavour's aft flight deck windows. Thuot is standing on the Remote Manipulator System end effector platform, attempting to attach a capture bar to the free floating communications satellite.

Edward H. White II drifts in the zero gravity of space in 1965, having become the first American to step outside his spacecraft and to walk in space. His experience lasted for a little over 20 minutes, and during his orbital stroll he travelled a distance of some 6,500 miles.

Against a backdrop of clouds 130 nautical miles (about 241 kilometres) below, astronaut Mark C. Lee floats freely without tethers as he tests the Simplified Aid for EVA Rescue (SAFER) system in 1994.

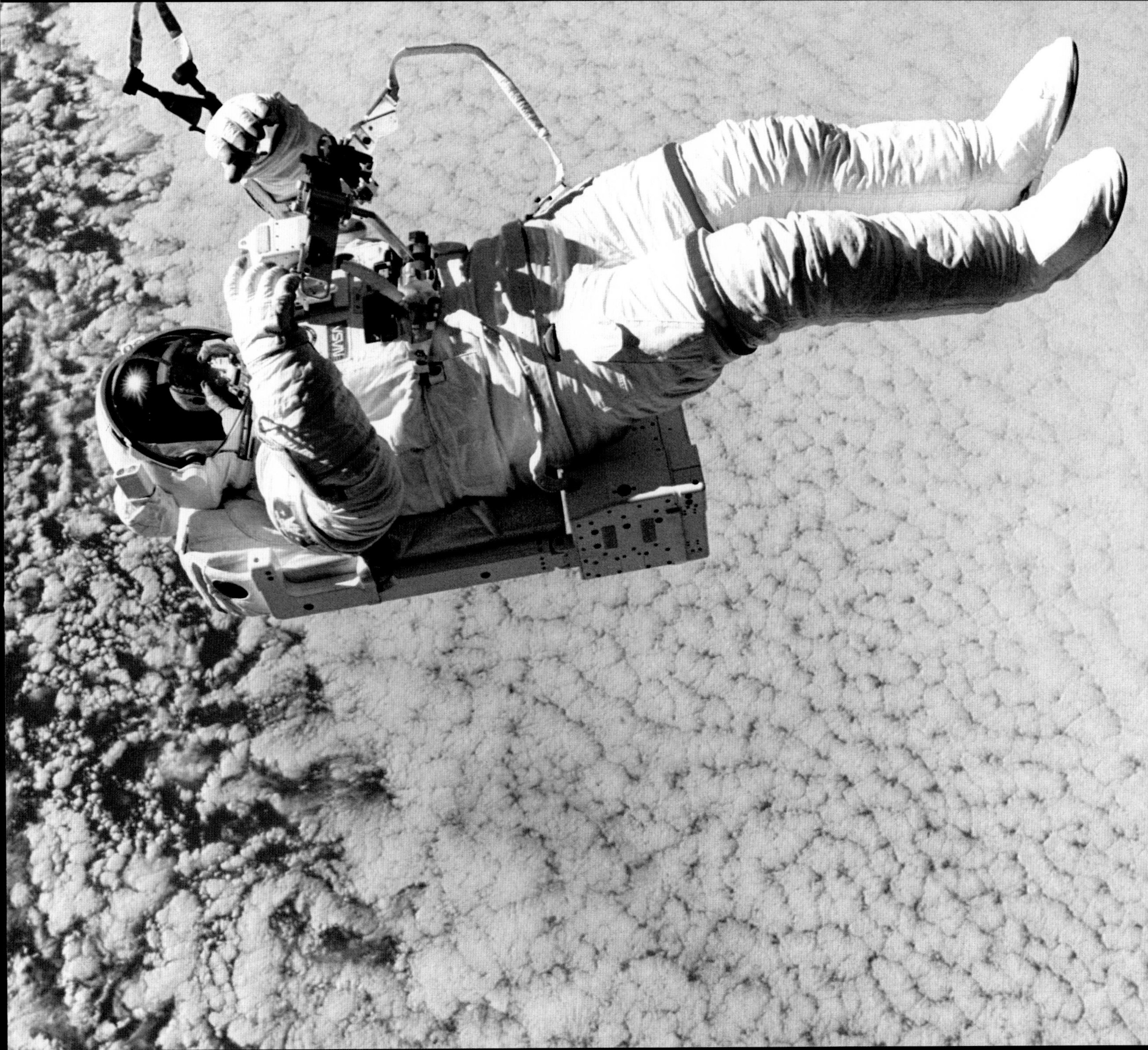

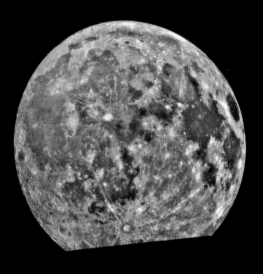

This false-colour photograph is a composite of 15 images of the Moon taken through three colour filters by the Galileo spacecraft's solid-state imaging system, during the spacecraft's passage through the Earth-Moon system on December 8, 1992. The false-colour processing used to create this lunar image allows the surface soil composition to be ascertained: areas appearing red generally correspond to the lunar highlands, while blue to orange shades indicate the ancient volcanic lava flow of a mare, or lunar sea. Bluer mare areas contain more titanium than do the orange regions.

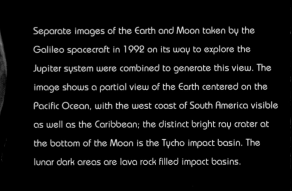

Separate images of the Earth and Moon taken by the Galileo spacecraft in 1992 on its way to explore the Jupiter system were combined to generate this view. The image shows a partial view of the Earth centered on the Pacific Ocean, with the west coast of South America visible as well as the Caribbean; the distinct bright ray crater at the bottom of the Moon is the Tycho impact basin. The lunar dark areas are lava rock filled impact basins.

Satellites need a way to generate power for months, even years, to run their on-board systems, and most Earth-orbiting spacecraft, such as the Hubble Space Telescope, rely on solar cells to recharge their onboard batteries. While solar panels have to be lightweight and flexible enough to fit inside a relatively small launch vehicle, they are also fragile and easily damaged, and Hubble suffered from this problem and needed a repair. Here astronaut Kathy Thornton releases the old panel into low-Earth orbit during the first Hubble Space Telescope Servicing Mission in 1993. Earth's gravitation pulled the jettisoned panel toward Earth's atmosphere, where it entered and ultimately burned up.

Fish-eye view of the Space Shuttle Atlantis as seen from the Russian Mir space station during the STS-71 mission in 1995.

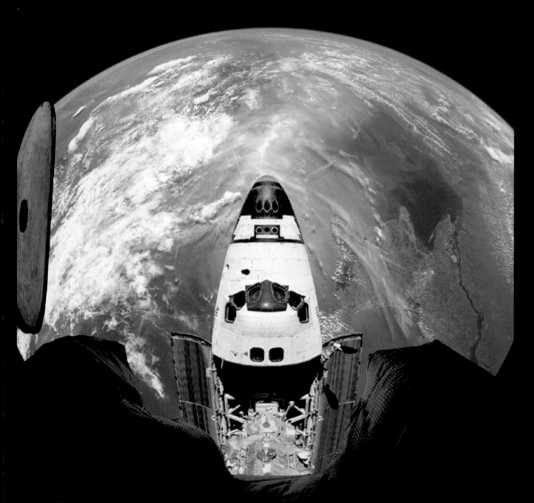

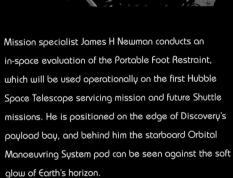

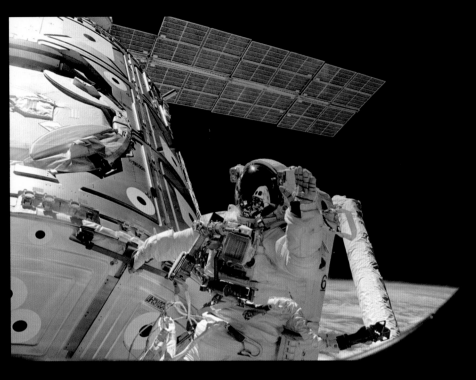

STS-88 mission specialist James Newman, holding on to a handrail, waves back at the camera during the first of three Extravehicular Activities (EVAs) performed during the shuttle's mission in 1993. The Orbiter itself can be seen reflected in his visor.

Mission specialist James H Newman conducts an in-space evaluation of the Portable Foot Restraint, which will be used operationally on the first Hubble Space Telescope servicing mission and future Shuttle missions. He is positioned on the edge of Discovery's payload bay, and behind him the starboard Orbital Manoeuvring System pod can be seen against the soft glow of Earth's horizon.

Earth serves as backdrop for astronaut Michael Gernhardt during his Extravehicular Activity in 1995. With Endeavour's forward section reflected in his visor, he is standing on a Manipulator Foot Restraint attached to the Remote Manipulator System. Unlike earlier spacewalking astronauts, Gernhardt was able to use an electronic cuff checklist, a prototype developed for the assembly of the International Space Station.

Edward H White II, pilot of the Gemini IV spacecraft, floats in the zero gravity of space with the Earth behind him as a backdrop. White's historic spacewalk in 1965 was performed during the third revolution of the Gemini IV spacecraft and it was the first time that an American had stepped outside the confines of his spacecraft. In his right hand he carries a Hand-Held Self-Manoeuvring Unit to allow him to control his movements, while the visor of his helmet is gold plated to protect him from the unfiltered rays of the sun.

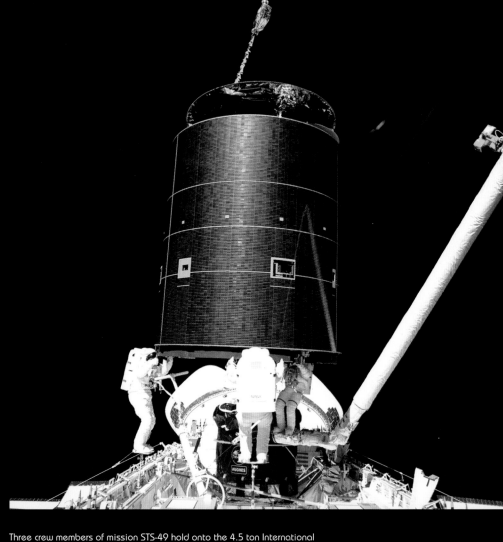

Three crew members of mission STS-49 hold onto the 4.5 ton International Telecommunications Organization Satellite (INTELSAT) VI after a six-handed 'capture' made minutes earlier during the mission's third Extravehicular Activity in 1992. Earlier mission specialist Pierre J Thuot had made two unsuccessful grapple attempts using the Remote Manipulator System arm, and ground controllers and crew members agreed that a third attempt, using three mission specialists in the payload bay was required.

An aerial view of the STS-2 Columbia launch from Pad 39A at the Kennedy Space Centre, Florida in 1989, taken by astronaut John Young, who was at the time aboard NASA's Shuttle Training Aircraft.

EARTH ORBIT

STS-49 mission specialist Pierre Thuot perches on the end effector of the
Robot Arm in 1992, armed with the Intelsat VI capture bar. This would
be one of many attempts to 'grapple' the Intelsat VI satellite, which was
rendered inoperative when its Payload Assist Module motor failed to fire,
preventing it from reaching an operational altitude.

Astronaut Bruce McCandless II, STS-41B mission specialist, tests a Mobile

In a scene pictured from the aft windows onboard the space shuttle Discovery on mission STS-95 in 1998, the Moon is framed between the Orbiter's OMS pod and the Earth.

Mission Specialist Peter J K Wisoff (bottom), wearing an Extravehicular Mobility Unit, works with the antenna on the European Retrievable Carrier while payload commander G David Low, on the Remote Manipulator System robot arm, hovers above. The two astronauts were conducting Detailed Test Objective procedures in the payload bay of Endeavour in 1993.

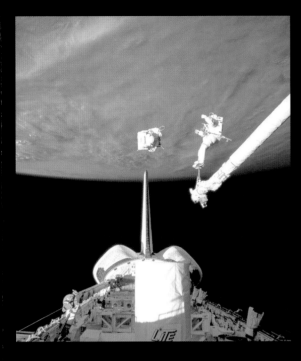

Astronauts Carl J Meade and Mark C Lee (red stripe on
suit) test the new Simplified Aid for EVA Rescue (SAFER)
system some 130 nautical miles (about 241 kilometres)
above Earth in 1994. They were actually performing an
in-space rehearsal or demonstration of a contingency
rescue using hardware previously untried in space.
Meade, who here wears the small back-pack unit with
its complementary chest-mounted control unit, and
Lee anchored to the Space Shuttle Discovery's Remote
Manipulator System robot arm, took turns using the
SAFER hardware during their shared space walk.

EARTH ORBIT

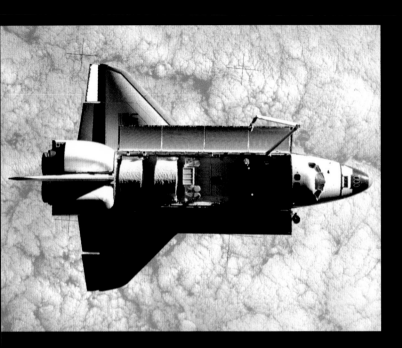

Full view of the Space Shuttle Challenger in space, taken by
the Space Pallet Satellite in 1993. Visible in the payload
bay are the protective cradles for the Palapa-B and
Telesat F communications satellites, the pallet for the NASA
Office of Space and Terrestrial Applications, the Remote
Manipulator System robot arm in the shape of the numeral
seven and the KU- band antenna. A number of GetAway
Special canisters are also visible along the port side.

Seen against a blue and white Earth, mission specialist and
payload commander G David Low and mission specialist
Peter J K Wisoff, both wearing Extravehicular Mobility Units,
simulate handling of large components in space in 1993.
This particular task was rehearsed with eyes toward the
servicing of the Hubble Space Telescope or the assembly
and maintenance of the International Space Station.

The Space Shuttle Atlantis seen
shortly after departing from the Mir
Russian Space Station in 1995. This
image was taken during the STS-71
mission by cosmonauts aboard their
Soyuz TM transport vehicle.

PART ONE: LOOKING OUT

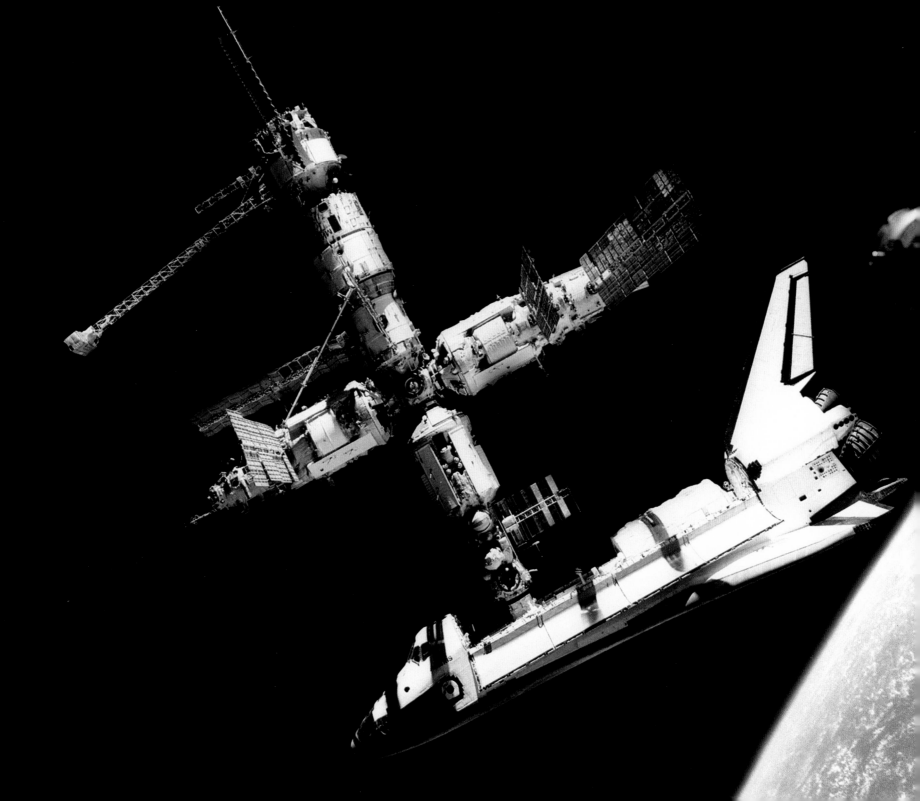

Astronaut Mark C Lee tests
out the Simplified Aid for EVA
Rescue (SAFER) system in 1994,
framed by a backdrop of a
cloud-covered Earth.

Disaster struck the space programme on January 28, 1986, when the Shuttle Challenger and her seven-member crew were lost when a ruptured O-ring in the right solid rocket booster caused an explosion soon after launch. Taken just a few seconds after the accident, this photograph shows the Space Shuttle main engines and solid rocket booster exhaust plumes entwined around a ball of gas from the external tank.

The Orbiter Discovery pictured on its approach to the Mir Russian Space Station in 1998. Visible in the payload bay is the Spacehab module and Alpha Magnetic Spectrometer payload.

pilot David Scott inside the
Command and Service Module
Gumdrop on the fifth day of the
Apollo 9 Earth-orbital mission
in 1969.

PART ONE: LOOKING OUT

View of the Apollo 9 Lunar Module Spider, in a lunar landing configuration, as photographed from the Command and Service Module on the fifth day of the Apollo 9 Earth-orbital mission. The landing gear on the Lunar Module has been deployed.

Taken with a handheld 70mm camera from a rendezvous window of America's Apollo spacecraft during the Apollo-Soyuz Test Project mission in 1975, this image shows the Soviet Soyuz spacecraft, whose docking system was specially designed to interface with the docking system on the Apollo's Docking Module. The goals of ASTP were to test the ability of American and Soviet spacecraft to rendezvous and dock in space and to open the doors to possible international rescue missions and future collaboration on manned spaceflights. The Soyuz and Apollo crafts launched from Baikonur and the Kennedy Space Center respectively, on July 15, 1975, and the two spacecraft successfully completed the rendezvous and docking on July 17.

Above: the Skylab Orbital Workshop (OWS) pictured from the Skylab IV Command and Service Module during the final fly-around by the CSM before returning home. Sixty-three seconds after launch on May 14, 1973 the micrometeor shield on the OWS experienced a failure that ripped it away and damaged the tie-downs that secured one of the solar array systems. Without the micrometeoroid shield temperatures inside the OWS rose to 52°C Centigrade (126°F), and the gold 'parasol,' seen in the picture, was designed to be fitted as a replacement, being deployed by the Skylab I crew. The repair ultimately enabled the OWS to fulfil all its mission objectives.

The International Space Station (ISS) moves away from the Space Shuttle Discovery. The US-built Unity node (top) and the Russian-built Zarya or FGB module (with the solar array panels deployed) were joined during a December 1998 mission.

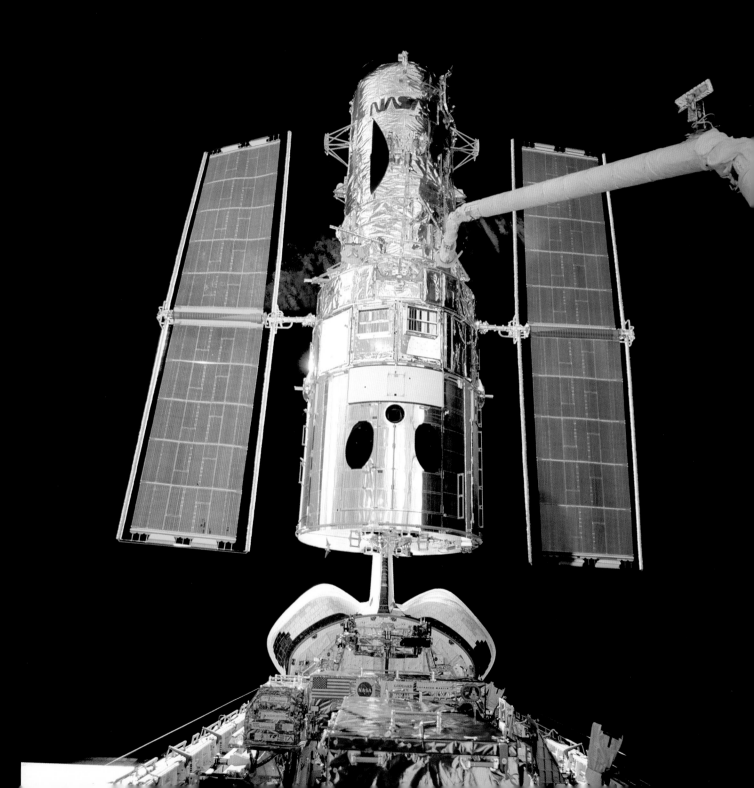

Attached to the robot arm of the shuttle, the Hubble Space Telescope is un-berthed and lifted up into the sunlight during the second mission to service the satellite in 1997.

The Space Shuttle Atlantis docked to the Kristall module of the Russian Mir Space Station in June 1995. The Space Shuttle/Mir combination, which was the largest space platform ever assembled, is shown overflying the Lake Baikal region of Russia.

The Russian-built FGB module, also called Zarya, nears the Space Shuttle Endeavour and the US-built Unity node, which can be seen in the foreground, in 1993. Inside Endeavour's cabin, the STS-88 crew ready the Remote Manipulator System, which will allow them to complete a successful rendezvous and to capture Zarya.

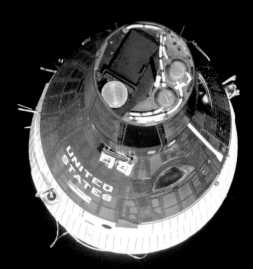

NASA successfully completed its first rendezvous mission with two Gemini spacecraft – Gemini VII and Gemini VI – in December 1965. This photograph, taken by Gemini VII crew members Frank Lovell and Frank Borman, shows Gemini VI in orbit 257 kilometres (160 miles) above Earth. The main purpose of Gemini VI, crewed by astronauts Walter Schirra and Thomas Stafford, was the meeting with Gemini VII, which was in turn studying the effects of long-duration space flight (up to 14 days) on a two-man crew.

Photographed on December 15, 1965, the Gemini VII spacecraft is seen from the hatch window of the Gemini VI spacecraft during rendezvous manoeuvres, the two craft staying at a distance of approximately nine feet apart.

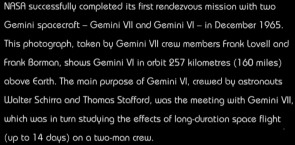

The Agena Target Vehicle as seen from the Gemini VIII spacecraft during rendezvous in 1966. This was the first time that two spacecraft had successfully docked, which was a critical milestone if a mission to the Moon was to become a reality.

The Hubble Space Telescope seen after being deployed following NASA's second servicing mission in 1997. The open aperture door of the telescope can be clearly seen.

The Augmented Target Docking Adapter as seen from the Gemini IX spacecraft in 1966 during one of the three rendezvous that took place between the craft in space. Although they were just over 66 feet apart at times, the docking adapter protective cover on the ATDA failed to fully separate and prevented the spacecraft from actually docking. The appearance of the ATDA led the Gemini crew to describe it at the time as an 'angry alligator.'

The damage to the ill-fated Apollo 13 Service Module can be clearly seen in this picture taken from the Lunar Module/Command Module following SM jettisoning in 1970. An entire SM panel had been blown away by the apparent explosion of an oxygen tank. The damage to the SM caused the Apollo 13 crewmen to use the Lunar Module as a 'lifeboat.' The Lunar Module 'Aquarius' was jettisoned just prior to Earth re-entry by the Command Module 'Odyssey.'

A 35mm camera was used to photograph sunlight over a cloud-covered Earth surface by STS-29 crew members onboard Discovery in March 1989.

Astronaut Robert L Gibson, STS-71 mission commander, shakes the hand of cosmonaut Vladimir N Dezhurov, Mir-18 commander. Their historic handshake took place in June 1995, two-and-a-half weeks prior to the twentieth anniversary of a similar in-space greeting between cosmonauts and astronauts participating in the Apollo-Soyuz Test Project.

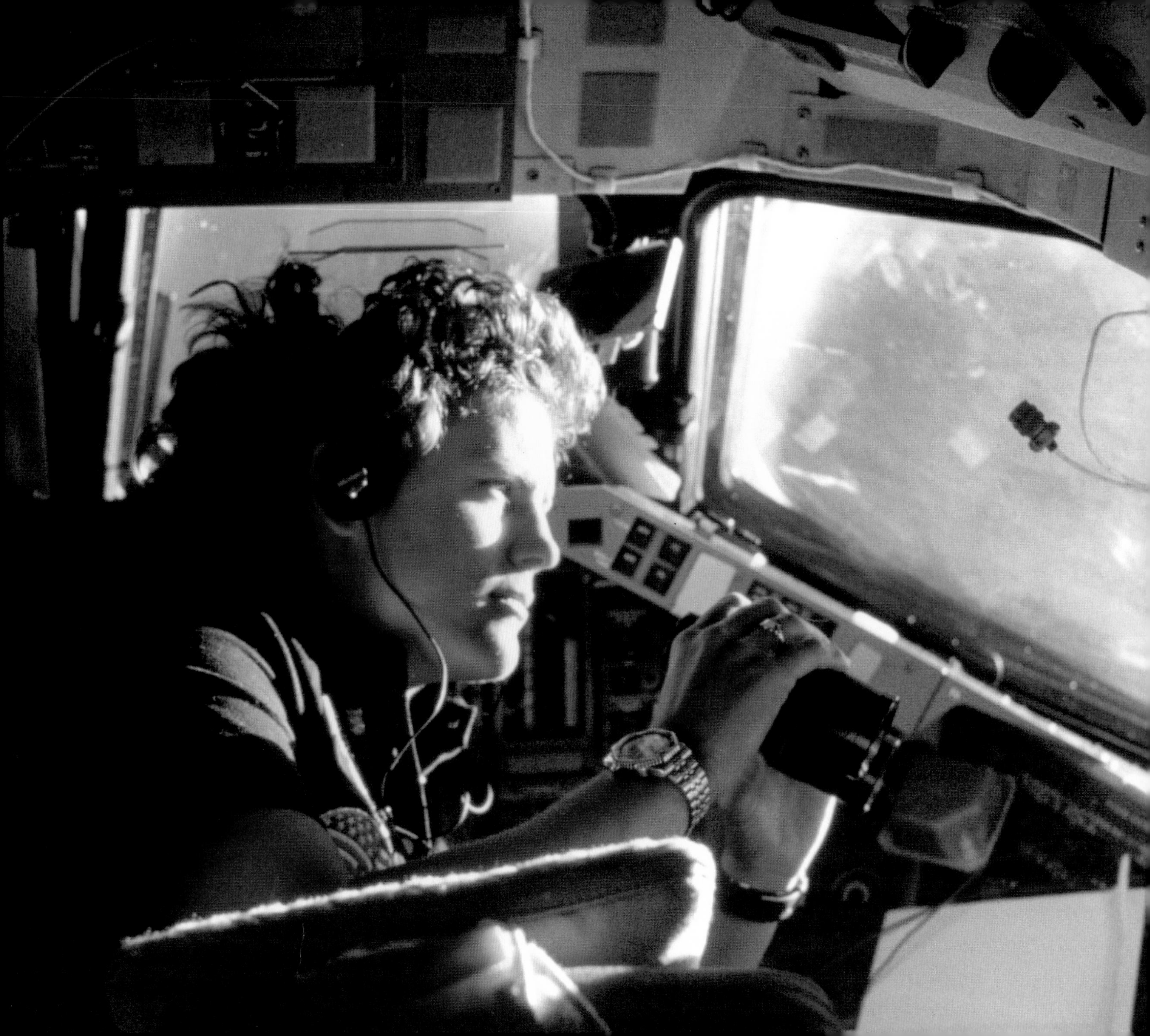

binoculars to take a closer look at
Earth through Challenger's forward
cabin windows in 1984.

spacecraft in 1965. The objective of the mission was to
evaluate and to test the effects of four days in space
on the crew, equipment and control systems. During the
mission White successfully accomplished the first
US space walk.

THE PLANETS

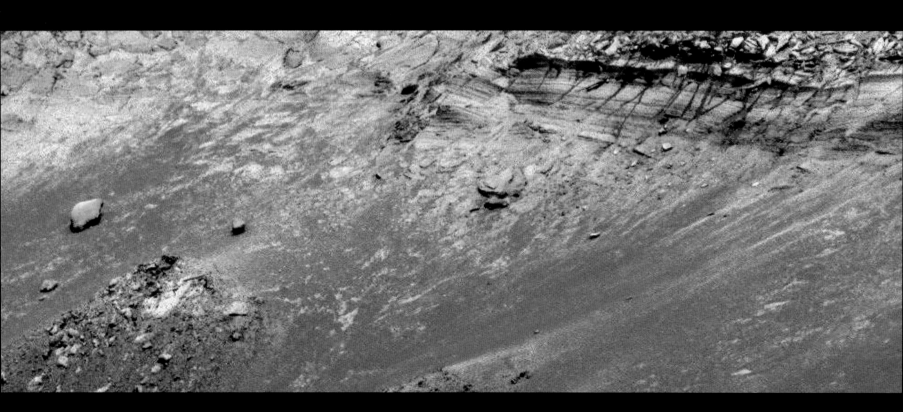

This false-colour image shows visible mineral changes between the materials
that make up the rim of the impact crater on Mars known as Endurance.
The image was taken by the panoramic camera on the Mars Exploration
Rover Opportunity using all 13 colour filters: the cyan blue denotes basalts,

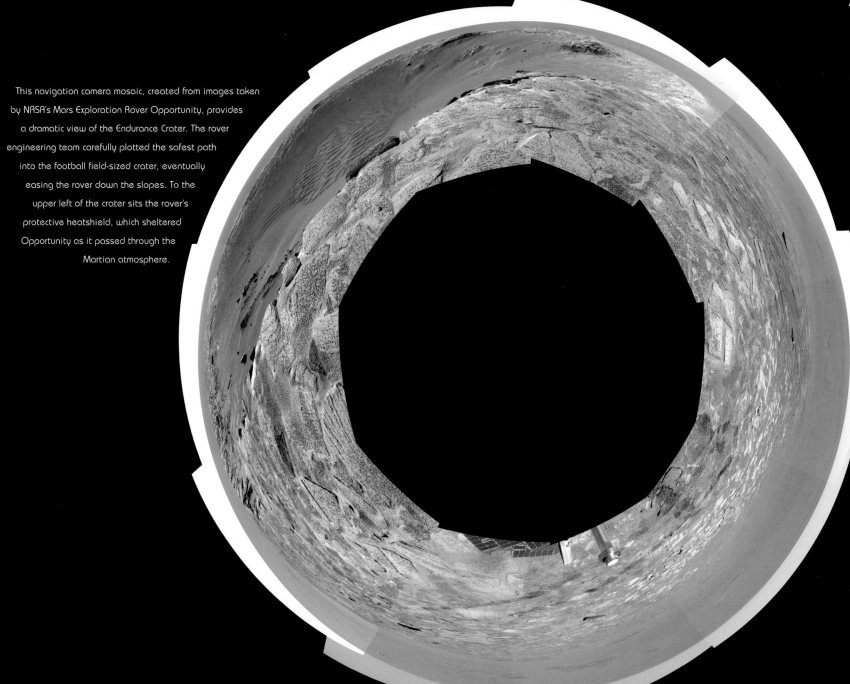

This navigation camera mosaic, created from images taken by NASA's Mars Exploration Rover Opportunity, provides a dramatic view of the Endurance Crater. The rover engineering team carefully plotted the safest path into the football field-sized crater, eventually easing the rover down the slopes. To the upper left of the crater sits the rover's protective heatshield, which sheltered Opportunity as it passed through the Martian atmosphere.

This mosaic was produced from pictures taken with the microscopic imager on NASA's Mars Exploration Rover Opportunity. It shows one of a series of holes ground by the rover's rock abrasion tool in Endurance Crater. This time Opportunity stretched its arm, or instrument deployment device, out to a target called Kettlestone. Grinding for just over two hours, Opportunity successfully created a hole 4.5 centimetres (1¾ inches) in diameter and 4.17 millimetres (1/8 inch) deep.

This image mosaic, taken by the navigation camera on NASA's Mars Exploration Rover Opportunity, shows the impact crater known as Endurance. The rover traversed the rim of the crater looking for clues to the crater's formation as well as a suitable entry point to allow it to go inside.

The Mars Exploration Rover Opportunity used its panoramic camera to capture this false-colour image of the interior of Endurance Crater on the rover's 188th day on the surface of Mars (August 4, 2004). The image data were relayed to Earth by the European Space Agency's Mars Express orbiter, and the image was generated from separate frames using the camera's 750-, 530- and 480-nanometer filters.

Three views of the full disk of Jupiter's volcanic moon, Io, each shown in natural and enhanced colour. These three views, taken by Galileo in late June 1996, show about 75 percent of Io's surface. Comparisons of these images to those taken by the Voyager spacecraft 17 years earlier, have revealed that many changes have occurred on Io, with about a dozen areas at least as large as the US state of Connecticut having been resurfaced.

NASA's Galileo spacecraft acquired its highest resolution images of Jupiter's moon, Io, on July 3, 1999 during its closest pass to Io since leaving the Earth in late 1995. This colour mosaic uses the near-infrared, green and violet filters (slightly more than the visible range) of the spacecraft's camera and approximates what the human eye would see. Most of Io's surface has pastel colours, punctuated by black, brown, green, orange, and red areas that denote active volcanoes.

As the Mars Exploration Rover Opportunity crept farther into Endurance Crater, the dune field on the crater floor appeared even more dramatic. This false-colour image taken by the rover's panoramic camera shows that the dune crests accumulated more dust than the flanks of the dunes and the flat surfaces between them. Also evident is a blue tint on the flat surfaces as compared to the dune flanks. This results from the presence of hematite-containing spherules that accumulate on the flat surfaces. Sinuous tendrils of sand less than 1 metre (3$^1/_3$ feet) high extend from the main dune field toward the rover. Scientists wanted to send the rover down one of these tendrils in an effort to learn more about the characteristics of the dunes, but the surface was too slippery. The dunes themselves are a common feature across the surface of Mars.

This true colour mosaic of Jupiter was constructed from images taken by the narrow angle camera on board NASA's Cassini spacecraft on December 29, 2000, during its closest approach to the giant planet at a distance of approximately 10 million kilometers (about 6 million miles). It is the most detailed global colour portrait of Jupiter ever produced, with the smallest visible features approximately 60 kilometers (37 miles) across. The mosaic is composed of 27 images: nine images were required to cover the entire planet in a tic-tac-toe pattern, and each of those locations was imaged in red, green, and blue to provide true colour. Although Cassini's camera can see more colours than humans can, Jupiter's colours in this view look very close to the way the human eye would see them.

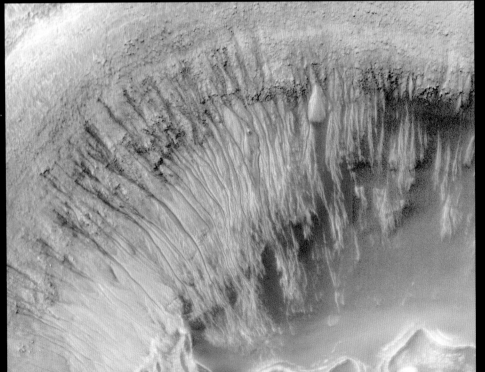

This approximate true-colour image taken by the Mars Exploration Rover Spirit shows a rock outcrop dubbed 'Longhorn' and, behind it, the sweeping plains of Gusev Crater. On the horizon, the rim of Gusev Crater is clearly visible. The image consists of four frames taken by the rover's panoramic camera.

Newton Crater is a large basin approximately 287 kilometres (178 miles) across on Mars, which was formed by an asteroid impact that probably occurred more than 3 billion years ago. The picture highlights the north wall of a specific, smaller crater located in Newton Crater's southwestern quarter. This has many narrow gullies eroded into it, which may have been created by running water.

This approximate true-colour rendering from the Mars Exploration Rover Spirit shows a set of darker rocks that are believed to be basaltic, or volcanic, in composition, because their spectral properties match those of other basaltic rocks studied in Gusev Crater.

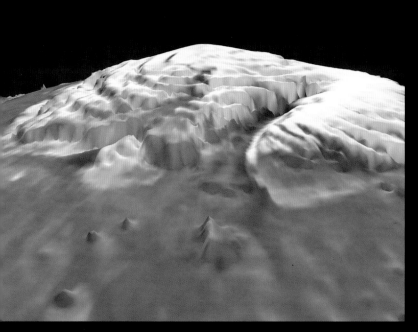

This three-dimensional image of the Mars north pole enabled scientists to estimate the volume of its water ice cap with unprecedented precision, and to study its surface variations and the heights of clouds in the region for the first time. Approximately 2.6 million laser pulse measurements were assembled into a topographic grid of the north pole with a spatial resolution of one kilometre

Mars Exploration Rover Spirit takes a good look around at its surroundings high above Gusev Crater. For the following eleven days the rover was out of reach as the Sun moved behind the Earth and Mars, blocking communications. Dominating the left side of this image, to the east, is the high point of the West Spur region of the Columbia Hills, where Spirit was exploring rock outcrops. On the right side are Spirit's tracks leading up the slope, and the dark areas show wheel tracks created when the rover slipped a little while negotiating the outcrops.

NASA's Mars Exploration Rover Opportunity examined a boulder about 1 metre (3¹/₃ feet) across called Wopmay before heading further east inside Endurance Crater. The frames combined into this false-colour view were taken by Opportunity's panoramic camera during the rover's 251st Martian day (October 7, 2004). The colouring accentuates iron-rich spherical concretions as bluish dots embedded in the rock and on the ground around it. The slope of the ground and loose surface material around the rock prevented Opportunity from getting firm enough footing to use its rock abrasion tool on Wopmay.

NASA's Mars Exploration Rover Spirit travelled more than 3 kilometres (2 miles) from its original landing site to reach the Columbia Hills pictured here. In this 360-degree view of the rolling Martian terrain, its wheel tracks can be seen approaching from the northwest (right side of image). Spirit's navigation camera took the images that make up this mosaic.

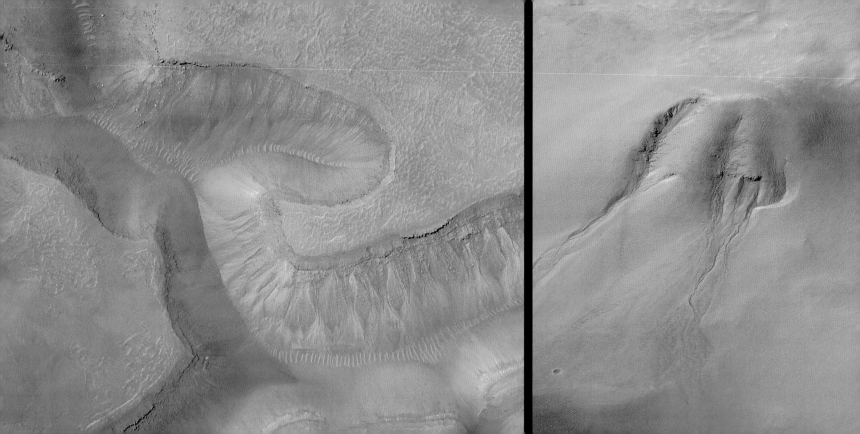

This self-portrait of the Mars Exploration Rover Opportunity comes courtesy of the Sun and the rover's front hazard-avoidance camera. The dramatic snapshot of Opportunity's shadow was taken as the rover moved into Endurance Crater.

This enhanced false-colour mosaic image from the Mars Exploration Rover
Spirit panoramic camera shows the view acquired after the rover drove
approximately 50.2 metres (165 feet) on the martian afternoon of sol 89
(April 3, 2004). The view shows the direction the rover is schedule to head
towards, and in the distance are the eastern-lying 'Columbia Hills.' This image
was assembled from images in the panoramic camera's near-infrared, green
and violet filters, and the colours have been exaggerated to enhance the
differences between cleaner and dustier rocks, and lighter and darker soils.

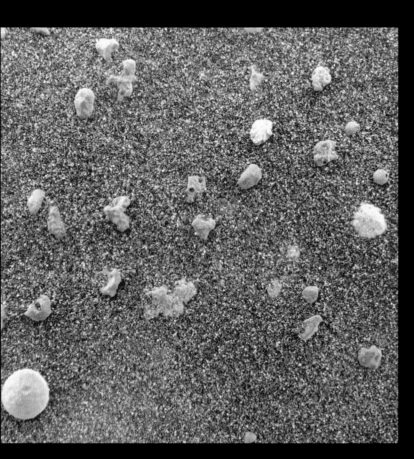

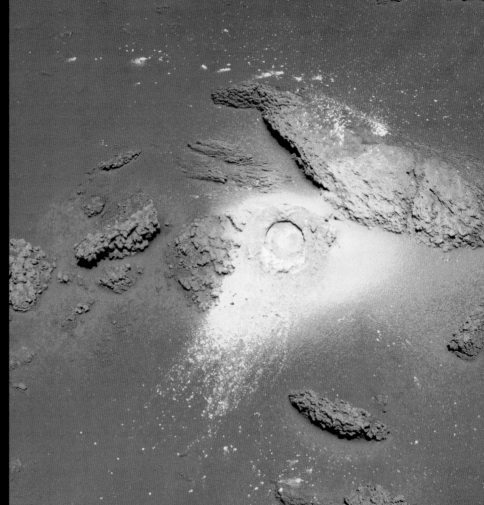

This magnified look at the martian soil near the

This image, taken by the panoramic camera on Mars Exploration Rover Spirit during the rover's trek through the Columbia Hills at Gusev Crater, shows the horizontally layered rock dubbed Tetl by NASA scientists, who were hoping to investigate it in more detail to determine whether the layering was volcanic or sedimentary in origin.

Extreme Ultraviolet Imaging Telescope (EIT) image of a huge, handle-shaped prominence on the surface of the Sun, taken in the 304 angstrom wavelength. Prominences are huge clouds of relatively cool dense plasma suspended in the Sun's hot, thin corona and, at times, they can erupt, escaping the Sun's atmosphere. The hottest areas in this image appear almost white, while the darker red areas indicate cooler temperatures.

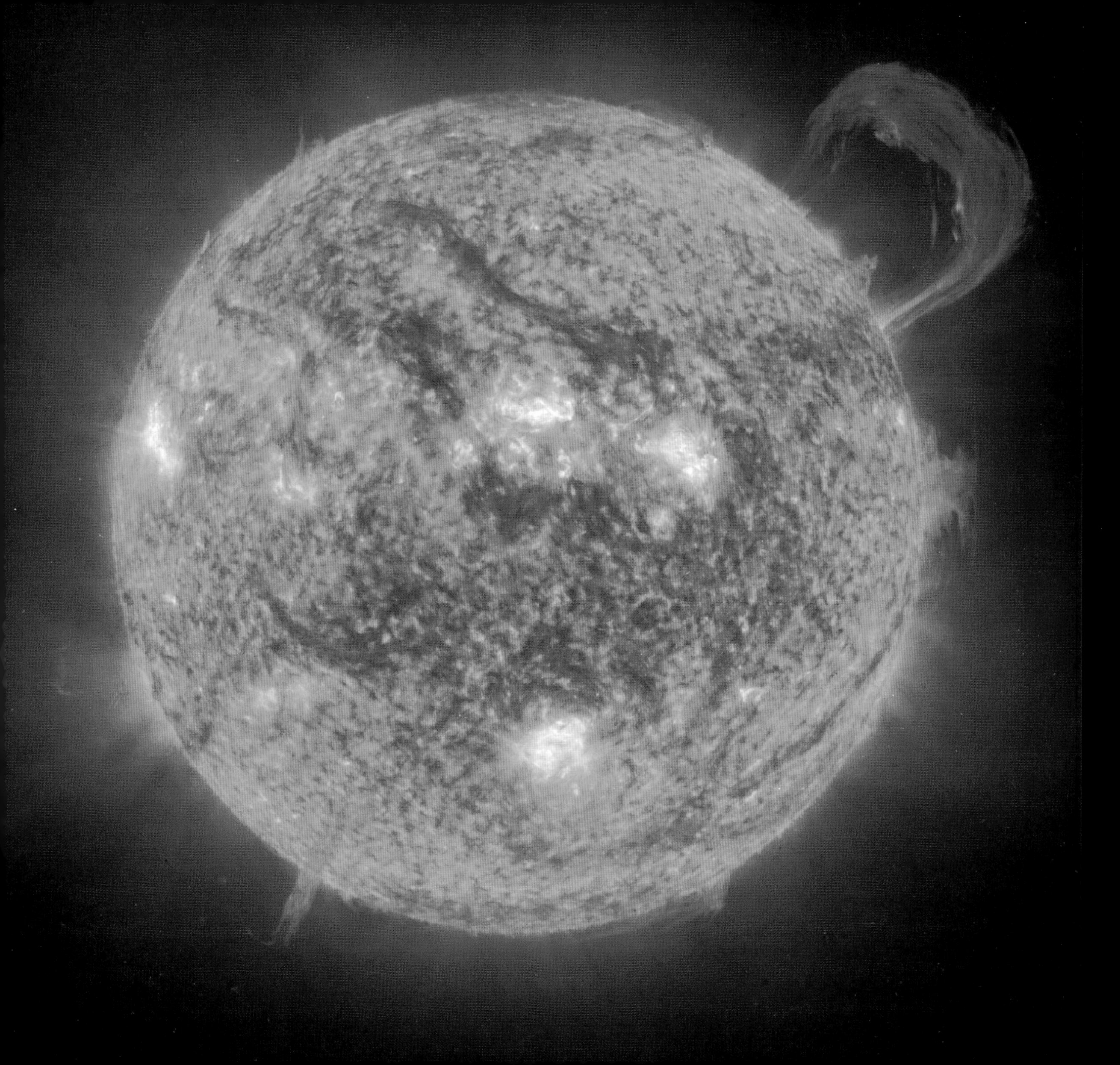

This picture of Neptune was produced from the last whole planet images taken through the green and orange filters on the Voyager II narrow angle camera in 1990. The images were taken at a range of 7.8 million kilometres (4½ million miles) from the planet, 4 days and 20 hours before closest approach. The picture shows the Great Dark Spot and its companion bright smudge; on the western horizon the fast moving bright feature called Scooter and the little dark spot are visible. Years later, when the Hubble telescope was focused on the planet, these atmospheric features had changed, indicating that Neptune's atmosphere is dynamic.

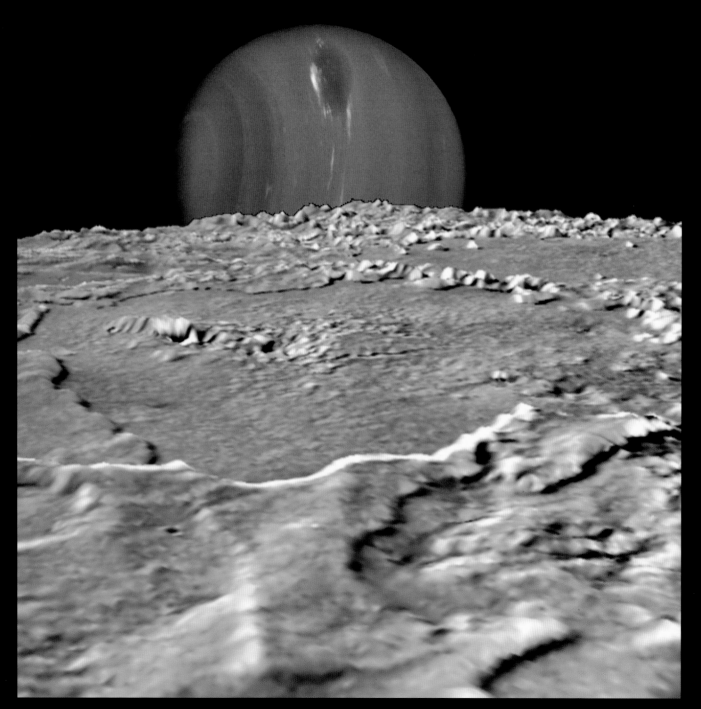

Composite view showing Neptune on its moon Triton's horizon. Neptune's south pole is to the left, while clearly visible in the planet's southern hemisphere is a Great Dark Spot, a large anticyclonic storm system located about 20 degrees South. The foreground is a computer generated view of Triton's surface as it would appear from a point approximately 45 km above the surface. This three-dimensional view was created from a Voyager image by using a two-dimensional photoclinometric model, in which relief has been exaggerated roughly 30-fold. In reality, would Neptune appear to be rising or setting? The answer is that it would do neither: due to the motion of Triton relative to Neptune, it would appear to move laterally along the horizon, eventually rising and setting at high latitudes.

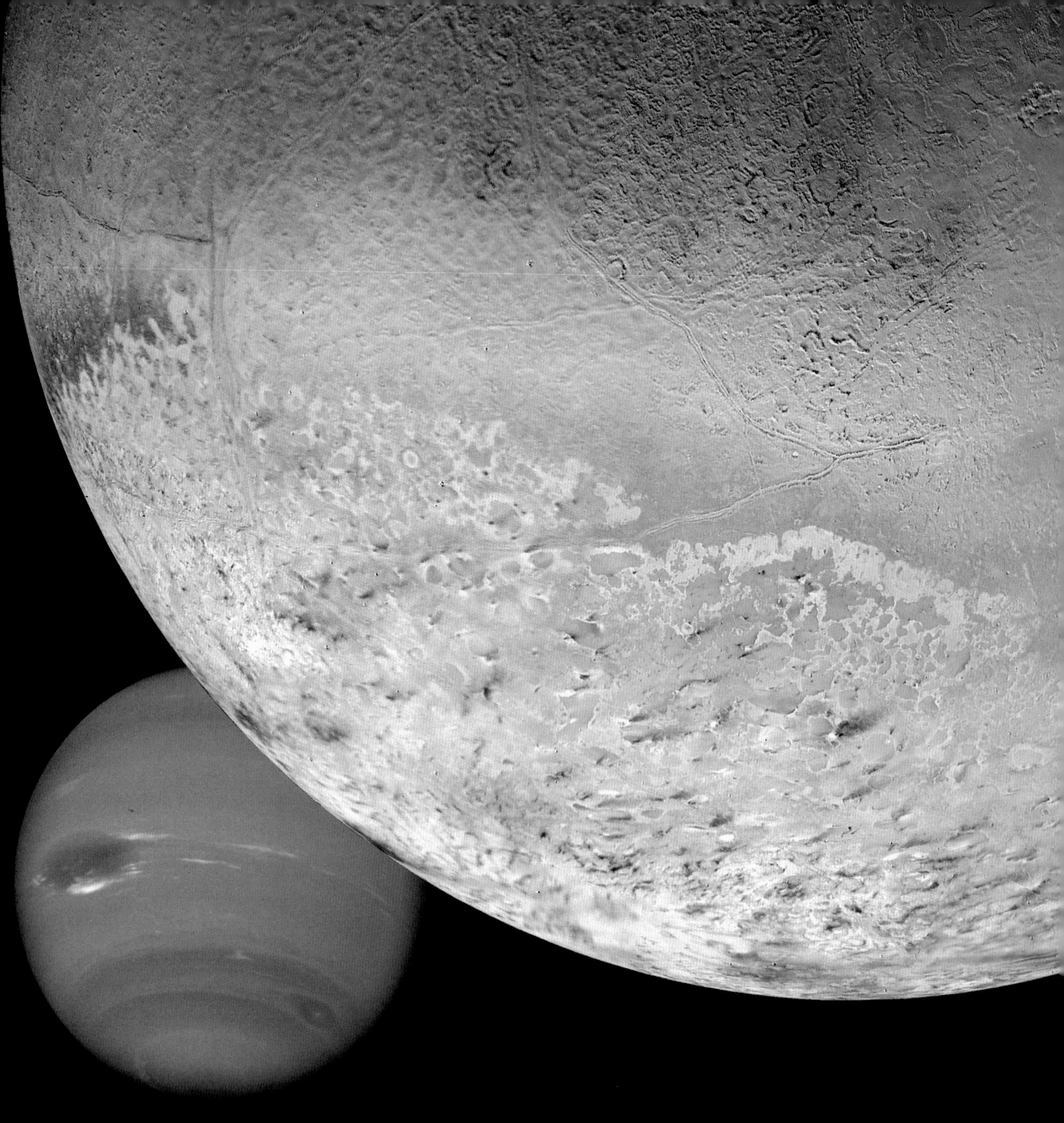

This computer generated montage shows Neptune as it would appear from a spacecraft approaching Triton, Neptune's largest moon at 2,706 kilometres (1,683 miles) in diameter. The wind and sublimation-eroded south polar cap of Triton is shown at the bottom of the Triton image, a cryovolcanic terrain at the upper right, and the enigmatic 'cantaloupe terrain' at the upper left. Triton's surface is mostly covered by nitrogen frost, mixed with traces of condensed methane, carbon dioxide, and carbon monoxide. The tenuous atmosphere of Triton, though only about one-hundredth of one percent of Earth's atmospheric density at the surface, is thick enough to produce wind-deposited streaks of dark and bright materials of unknown composition in the south polar cap region.

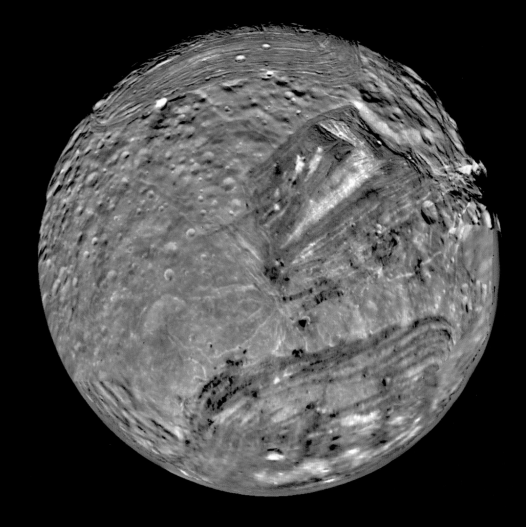

This image was returned by the Voyager 2 spacecraft on July 3, 1989, when it was 76 million kilometres (47 million miles) from Neptune. The planet and its largest satellite, Triton, were captured in the field of view of Voyager's narrow-angle camera through violet, clear and orange filters. Triton appears in the lower right corner at about 5 o'clock relative to Neptune.

Uranus moon Miranda is shown in a computer-assembled mosaic of images obtained by the Voyager II spacecraft. Miranda is the innermost and smallest of the five major Uranian satellites, just 480 kilometers (about 300 miles) in diameter. Nine images were combined to obtain this full-disc, south-polar view: the bulk of the photo comprises seven high-resolution images from the Voyager closest-approach sequence. Data from more distant, lower-resolution images were used to fill in gaps along the horizon.

THE PLANETS

The varying temperatures of Saturn's rings are depicted here in this false-colour image made from data obtained by the Cassini spacecraft's composite infrared spectrometer instrument. Red represents temperatures of about 110 Kelvin (-261°F), and blue 70 Kelvin (-333°F). Green is equivalent to 90 Kelvin (-298°F). Water freezes at 273 Kelvin (32°F). The spatial resolution of the ring portion of the image is 200 kilometres (124 miles).

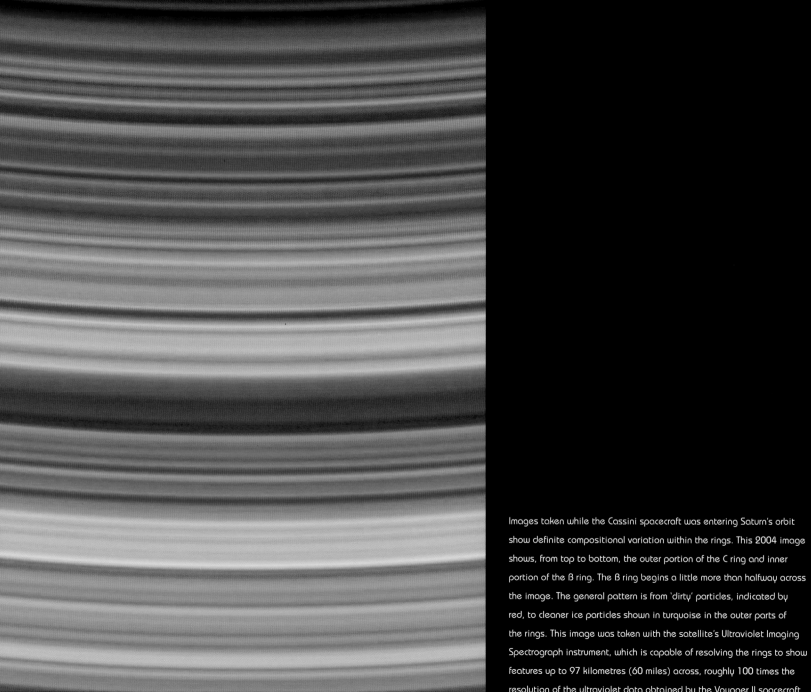

Images taken while the Cassini spacecraft was entering Saturn's orbit show definite compositional variation within the rings. This 2004 image shows, from top to bottom, the outer portion of the C ring and inner portion of the B ring. The B ring begins a little more than halfway across the image. The general pattern is from 'dirty' particles, indicated by red, to cleaner ice particles shown in turquoise in the outer parts of the rings. This image was taken with the satellite's Ultraviolet Imaging Spectrograph instrument, which is capable of resolving the rings to show features up to 97 kilometres (60 miles) across, roughly 100 times the resolution of the ultraviolet data obtained by the Voyager II spacecraft.

While cruising around Saturn, the Cassini orbiter captured a series of images that have been composed into the largest and most detailed global natural colour view of Saturn and its rings ever made. This grand mosaic consists of 126 images acquired in a tile-like fashion, covering one end of Saturn's rings to the other and the entire planet in between. The images were taken over the course of two hours on October 6, 2004, while Cassini was approximately 6.3 kilometres (3.9 million miles) from Saturn, and the smallest features seen here are just 38 kilometres (24 miles) across.

At first glance, Jupiter looks like it has a mild case of the measles. Five spots – one coloured white, one blue, and three black – are scattered across the upper half of the planet. Closer inspection by NASA's Hubble Space Telescope reveals that these spots are actually a rare alignment of three of Jupiter's largest moons – Io, Ganymede, and Callisto – across the planet's face.

In this image, the telltale signatures of this alignment are the shadows (the three black circles) cast by the moons. Io's shadow is located just above centre and to the left; Ganymede's is on the planet's left edge; and Callisto's is near the right edge. Only two of the moons, however, are visible in this image. Io is the white circle in the centre of the image, while Ganymede is the blue circle at upper right. Callisto is out of the image and to the right.

Nine days before it entered orbit in 2004, the Cassini spacecraft captured this exquisite natural colour view of Saturn's rings. The images that make up this composition were obtained from Cassini's vantage point beneath the ring plane with the craft's narrow angle camera, at a distance of 6.4 million kilometers (4 million miles) from Saturn. The brightest part of the rings, curving from the upper right to the lower left in the image, is the B ring. Colour variations in Saturn's rings had previously been seen in Voyager and Hubble Space Telescope images. Cassini's images show that colour variations in the rings are more pronounced in this viewing geometry than they are when seen from Earth.

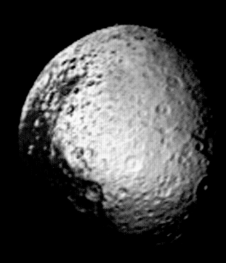

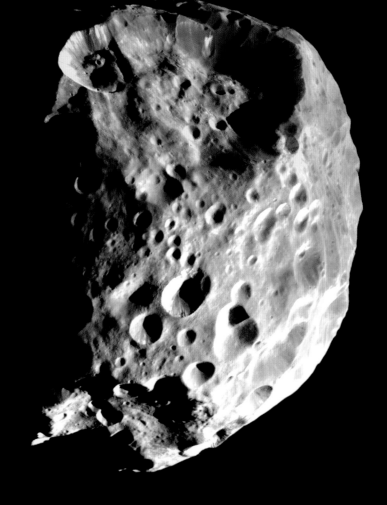

Saturn's outermost large moon, Iapetus, has a bright, heavily cratered icy terrain and a dark terrain, as shown in this Voyager II image taken on August 22, 1981. Amazingly, the dark material covers precisely the side of Iapetus that leads in the direction of orbital motion around Saturn (except for the poles), whereas the bright material occurs on the trailing hemisphere and at the poles. The bright terrain is made of dirty ice, and the dark terrain is surfaced by carbonaceous molecules, according to measurements made with Earth-based telescopes. Iapetus' dark hemisphere has been likened to tar or asphalt and is so dark that no details within this terrain were visible to Voyager II. The bright icy hemisphere meanwhile, likened to dirty snow, shows many large impact craters. The closest approach by Voyager II to Iapetus was a relatively distant 600,000 miles.

The true nature of Saturn's moon Phoebe is revealed in startling clarity in this mosaic of two images taken during the flyby made by Cassini on June 11, 2004. The image shows evidence for the emerging view that Phoebe may be an ice-rich body coated with a thin layer of dark material. Small bright craters in the image are probably fairly young features, and this phenomenon has been observed on other icy satellites, such as Ganymede at Jupiter. When meteorites slammed into the surface of Phoebe, the collisions excavated fresh, bright material – probably ice – underlying the surface layer. Further evidence for this can be seen on some crater walls where the darker material appears to have slid downwards, exposing more light-colored material. This spectacular view was obtained at a phase angle of 84 degrees, and from a distance of approximately 32,500 kilometers (20,200 miles).

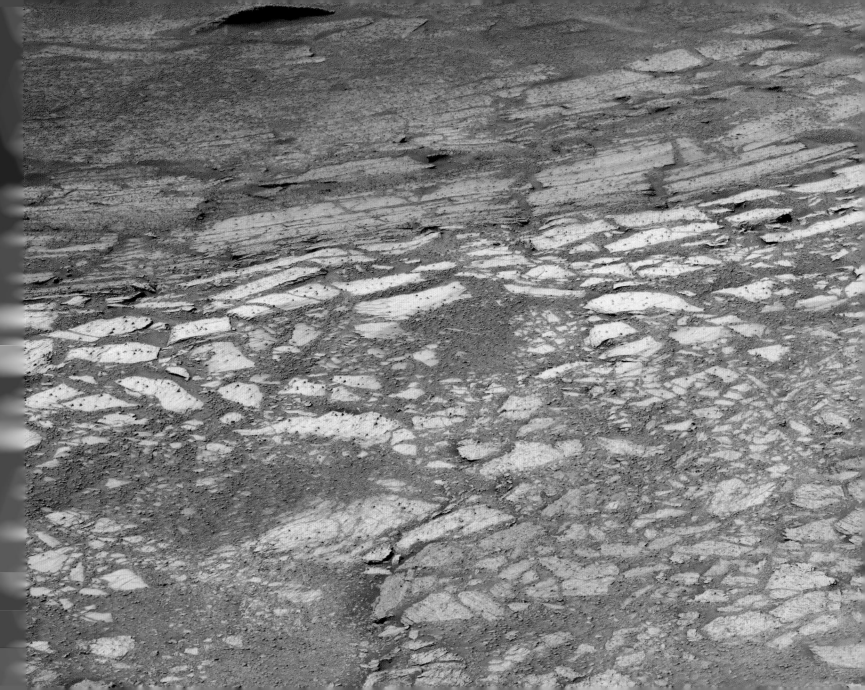

This false-colour image shows an area within Endurance Crater on Mars. The NASA rover was inspecting a hole it drilled into a flat rock dubbed 'Tennessee,' which scientists believed might have been made up of the same evaporite-rich materials as those found in nearby Eagle Crater. The overall geography inside Endurance proved to be more complex than scientists anticipated, with at least three distinct bands of rock visible in front of the rover.

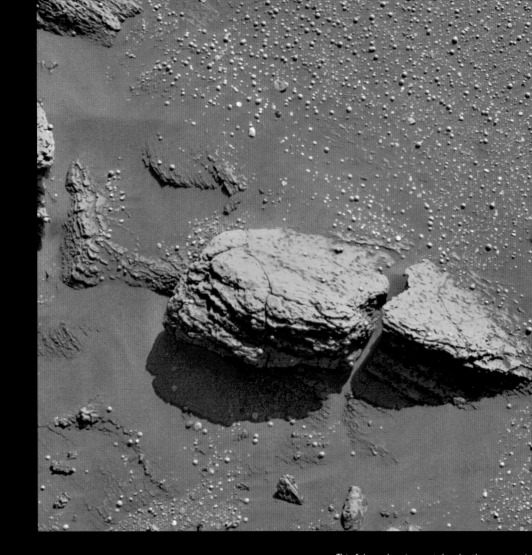

This false-colour image taken by the panoramic camera on board the Mars Exploration Rover Opportunity highlights the spherules that speckle the rock dubbed Stone Mountain. The colours in this picture were exaggerated or stretched to enhance the real difference in colour between Stone Mountain and its collection of

This black and white image of Jupiter's moon Callisto was
taken in 1979 by Voyager II from a range of about
1.1 million kilometres (675,000 miles), and the
picture has been enhanced to reveal detail in
the scene. Callisto exhibits some of the most
ancient terrain seen on any of the satellites.
Scientists think its surface is a mixture
of ice and rock dating back to the final
stages of planetary accretion (over 4
billion years ago) when the surface
was pockmarked by a torrential
bombardment of meteorites.
Younger craters show as bright
spots, probably because they
expose fresh ice and frost.

The pointed features shown here may only be a few centimetres high and less than 1 centimetre (0.4 inches) wide, but major scientific interest was generated by this 2004 image from the surface of Mars. Dubbed 'Razorback,' this chunk of rock sticks up at the edge of flat rocks in Endurance Crater. Based on their understanding of processes on Earth, scientists believe these features may have formed when fluids migrated through fractures, depositing minerals. Fracture-filling minerals would have formed veins composed of a harder material that eroded more slowly than the rock slabs. Examination of these features may further explain what they have to do with the history of water on Mars.

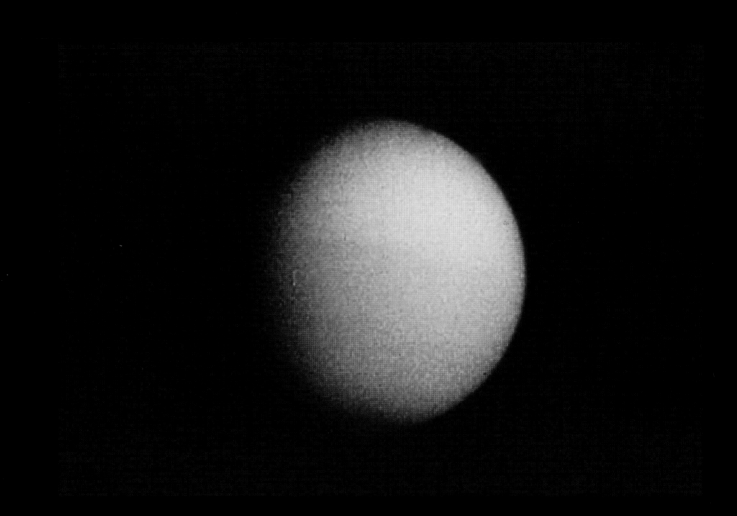

This false-colour image taken by the Mars Exploration Rover Spirit shows a group of darker rocks dubbed 'Toltecs.' The rocks are believed to be basaltic, or volcanic, in composition because their colour and spectral properties resemble those of basaltic rocks studied so far at Gusev Crater.

This image of Saturn's moon Titan, taken by a Voyager spacecraft, shows it to be completely shrouded by a thick atmosphere. The atmosphere is about 95 percent nitrogen, the remainder methane as well as other hydrocarbons and hydrogen cyanide.

THE PLANETS

This image of Ganymede, Jupiter's largest moon, was taken in 1979 by Voyager I from a distance of 2.6 million kilometers (1½ million miles). Ganymede has a bulk density of only approximately 2.0 g/cc, almost half that of our Moon, indicating that it is probably composed of a mixture of rock and ice. The features here, the large dark regions, in the northeast quadrant, and the white spots, resemble the mare and impact craters that are found on the Moon. Meanwhile the long white filaments resemble rays that are associated with impacts on the lunar surface. The various colours of different regions probably represent differing surface materials.

A global colour mosaic of Triton, taken in 1989 by Voyager II during its flyby of the Neptune system. Colour was synthesized by combining high-resolution images taken through orange, violet, and ultraviolet filters. These images were displayed as red, green, and blue images and combined to create this colour version. With a radius of 1,350 kilometres (839 miles), about 22 percent smaller than Earth's moon, Triton is by far the largest satellite of Neptune. It is one of only three objects in the Solar System known to have a nitrogen-dominated atmosphere (the others are Earth and Saturn's giant moon, Titan). Triton has the coldest surface known anywhere in the Solar System: 38 Kelvin (about -235°C or -391°F). It is so cold that most of Triton's nitrogen is condensed as frost, making it the only satellite in the Solar System known to have a surface made mainly of nitrogen ice

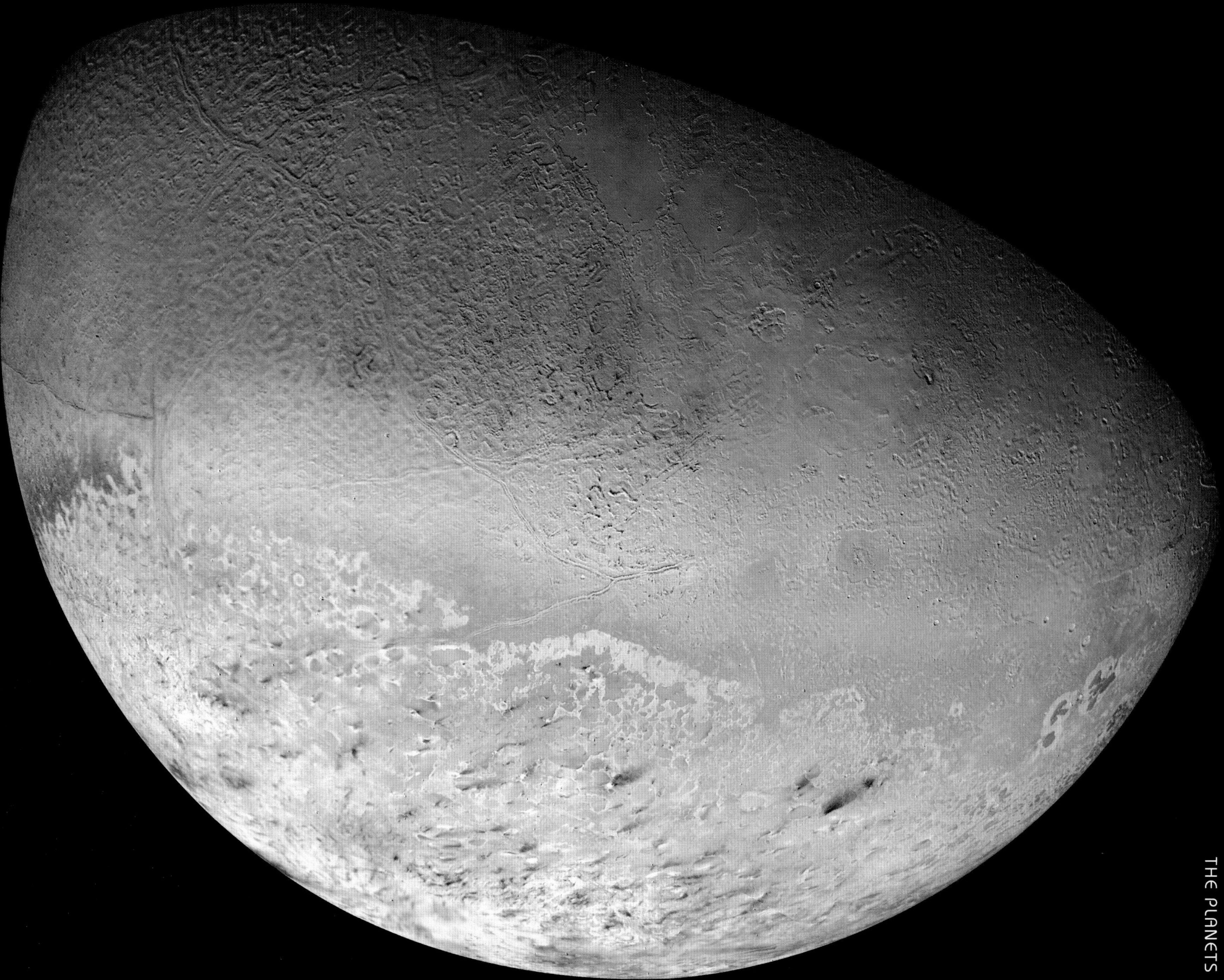

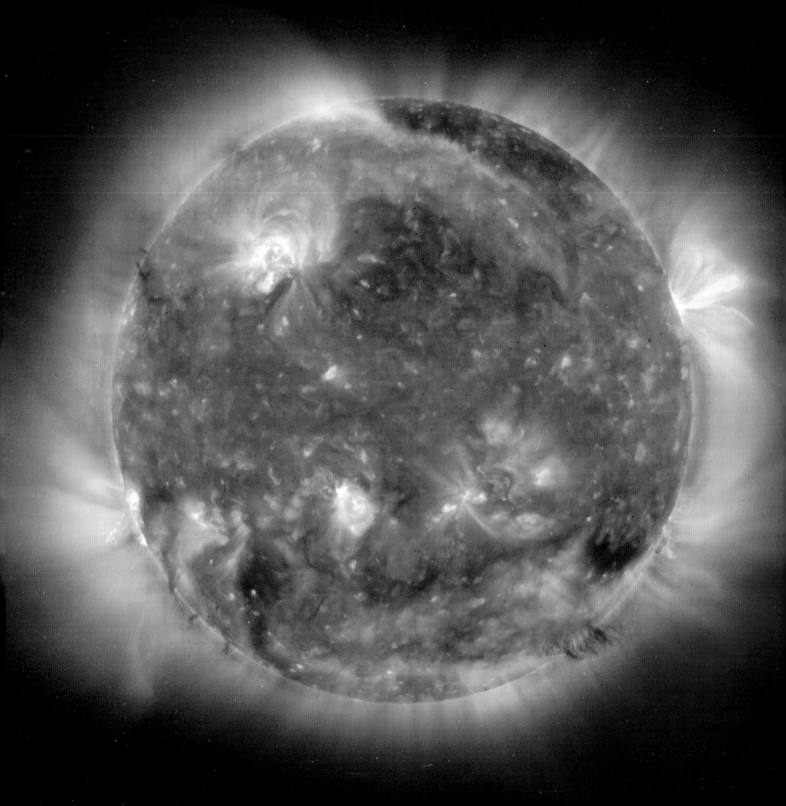

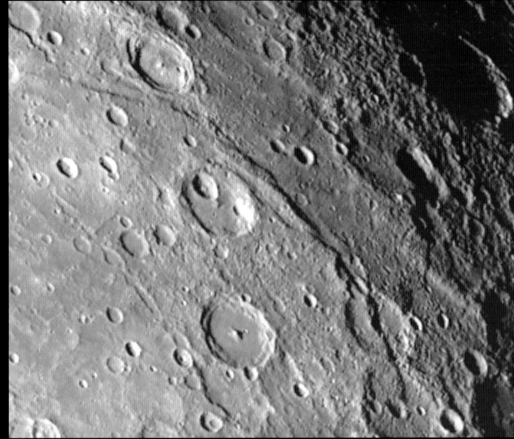

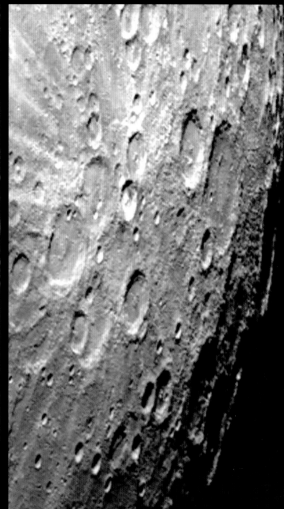

This composite image of the Sun combines Extreme Ultraviolet Imaging Telescope (EIT) images from three wavelengths (171, 195 and 284 angstrom) into one that reveals solar features unique to each wavelength. Since the EIT images are sent back to Earth in black and white, they are colour coded for easy identification. For this image, the nearly simultaneous images were each given a code (red, yellow and blue) and merged into one.

A scarp, or cliff, more than 300 kilometres (185 miles) long extends diagonally from upper left to lower right in this 1974 Mariner X picture of Mercury. Numerous similar structures have been discovered by Mariner X during the television sequences on the spacecraft's second flyby of the planet. These structures are believed to be formed by the compressive forces created by crustal shortening. The picture was taken from 64,500 kilometres (40,000 miles).

After passing Mercury the first time and making a trip around the Sun, Mariner X flew by Mercury for a second time, and this encounter brought the spacecraft in front of Mercury in the southern hemisphere. The Mariner X mission explored Venus in February 1974 on the way to three encounters with Mercury, in March and September 1974 and in March 1975. The spacecraft took more than 7,000 photos of Mercury, Venus, the Earth and the Moon.

BEYOND THE
SOLAR SYSTEM I

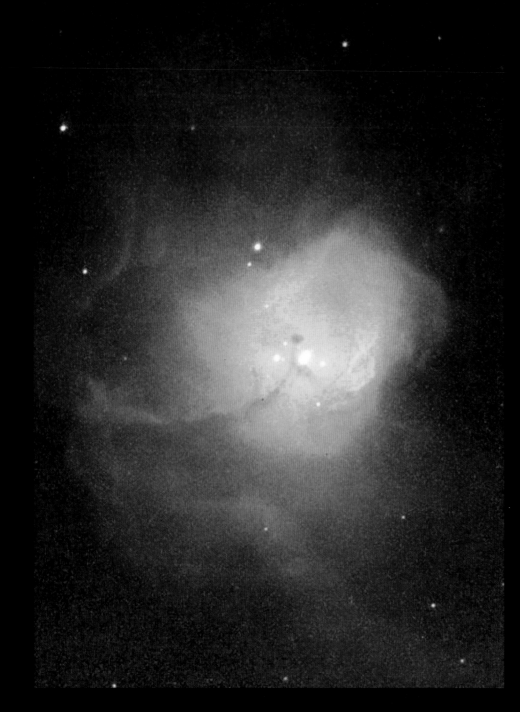

This stellar swarm is M80 (NGC 6093), one of the densest of the 147 known globular star clusters in the Milky Way Galaxy. Located about 28,000 light-.years from Earth, M80 contains hundreds of thousands of stars, all held together by their mutual gravitational attraction. Globular clusters are particularly useful for studying stellar evolution, since all of the stars in the cluster have the same age (about 15 billion years), but cover a range of stellar masses. Every star visible in this image is either more highly evolved than, or in a few rare cases more massive than, our own Sun. Especially obvious are the bright red giants, which are stars similar to the Sun in mass that are nearing the ends of their lives.

The Hubble Space Telescope took this 'family portrait' of young, ultra-bright stars nested in their embryonic cloud of glowing gases. The celestial maternity ward, called N81, is located 200,000 light- years away in the Small Magellanic Cloud (SMC), a small irregular satellite galaxy of our Milky Way. These are probably the youngest massive stars ever seen in the SMC. The nebula offers a unique opportunity for a close-up glimpse at the firestorm accompanying the birth of extremely massive stars, each blazing with the brilliance of 300,000 of our suns. Such galactic fireworks were much more common billions of years ago in the early universe, when most star formation took place. The picture was taken with Hubble's Wide Field and Planetary Camera 2.

A star 40 times more massive than the Sun, known as the Bubble Nebula, is blowing a giant bubble of material into space. The Hubble telescope captured the beefy star (lower centre), which is seen embedded in the bright blue bubble. The stellar powerhouse is so hot that it is quickly shedding material into space, while the dense gas surrounding the star is shaping the castoff material into a bubble. The bubble's surface is not smooth like a soap bubble's however: its rippled appearance is due to encounters with gases of different thickness. The nebula is 6 light-years wide and is expanding at 4 million miles per hour (7 million kilometre s per hour). The nebula is located 7,100 light-years from Earth in the constellation Cassiopeia.

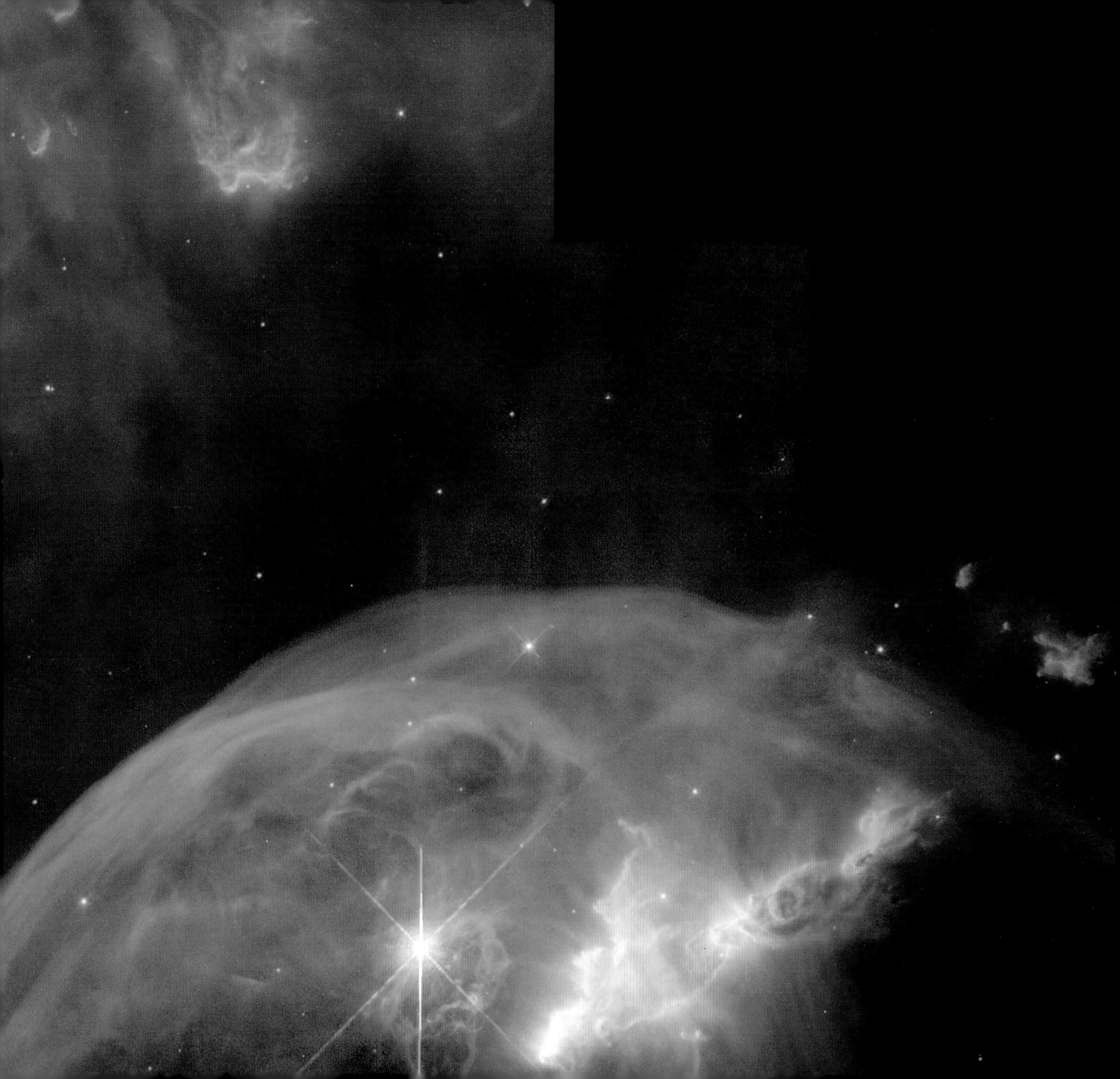

Weeks after NASA astronauts
repaired the Hubble Space
Telescope in December 1999,
the Hubble Heritage Project
took this picture of NGC 1999, a
nebula in the constellation Orion.
The Heritage astronomers, in
collaboration with scientists in
Texas and Ireland, used Hubble's
Wide Field Planetary Camera 2
(WFPC2) to get the colour image.

This is a Hubble image of a vast
nebula called NGC 604, which lies
in the neighbouring spiral galaxy
M33, 2.7 million light-years away
in the constellation Triangulum.
This is a site where new stars are
being born in a spiral arm of the
galaxy. Though such nebulae are
common in galaxies, this one is
especially large, nearly 1,500 light-
years across. At the heart of NGC
604 are over 200 hot stars, much
more massive than our Sun (15 to
60 solar masses). They heat the
gaseous walls of the nebula making
the gas fluoresce, and their light
also highlights the nebula's
three-dimensional shape, like a
lantern in a cavern. The image
was taken on January 17, 1995
with Hubble's WFPC2. Separate
exposures were taken in different
colours of light to study the physical
properties of the hot gas.

From its orbit around Earth, the Goddard Space Flight Center's Cosmic Background Explorer (COBE) captured this edge-on view of the Milky Way galaxy in infrared light, as part of its mission to test the big-bang theory of the creation of the universe. The satellite discovered that the cosmic background radiation had indeed been produced in the Big Bang just as scientists originally speculated. The satellite's data even discovered the primordial temperature and density fluctuations that eventually gave rise to the Milky Way and other large-scale objects found in space today.

Previously unseen details of a mysterious, complex structure within the Carina Nebula (NGC 3372) are revealed by this Hubble image of the Keyhole Nebula. The picture is a montage assembled from four different April 1999 telescope pointings with Hubble's Wide Field Planetary Camera 2, which used six different colour filters. The picture is dominated by a large, approximately circular feature, which is part of the Keyhole Nebula, named in the nineteenth century by Sir John Herschel. This region, about 8,000 light-years from Earth, is located adjacent to the famous explosive variable star Eta Carinae, which lies just outside the field of view toward the upper right.

This Hubble image of an expanding halo of light around a distant star, named V838 Monocerotis (V838 Mon), was obtained with the Advanced Camera for Surveys on February 8, 2004. The illumination of interstellar dust comes from the red supergiant star at the middle of the image, which gave off a flashbulb-like pulse of light two years previously. V838 Mon is located about 20,000 light-years away from Earth in the direction of the constellation Monoceros, placing the star at the outer edge of the Milky Way galaxy.

BEYOND THE SOLAR SYSTEM |

After the space shuttle
Discovery had redeployed
the Hubble Space Telescope
on completion of its second
servicing mission, the silvery
telescope, with its aperture
door open, is seen in sharp
contrast against the velvety
blackness of space.

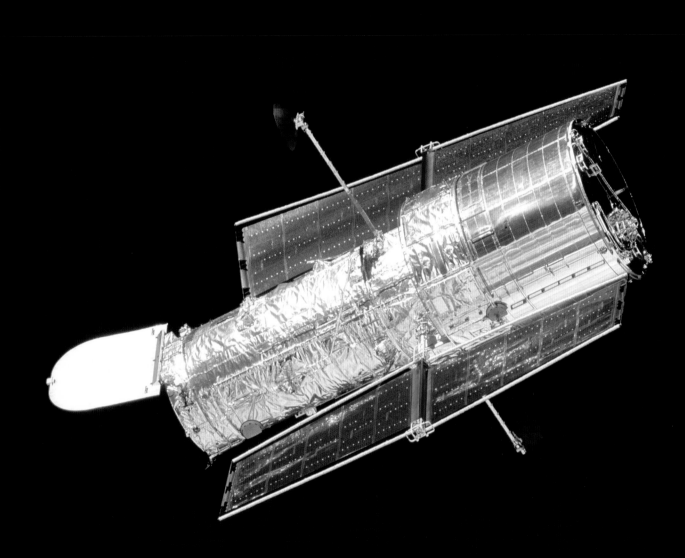

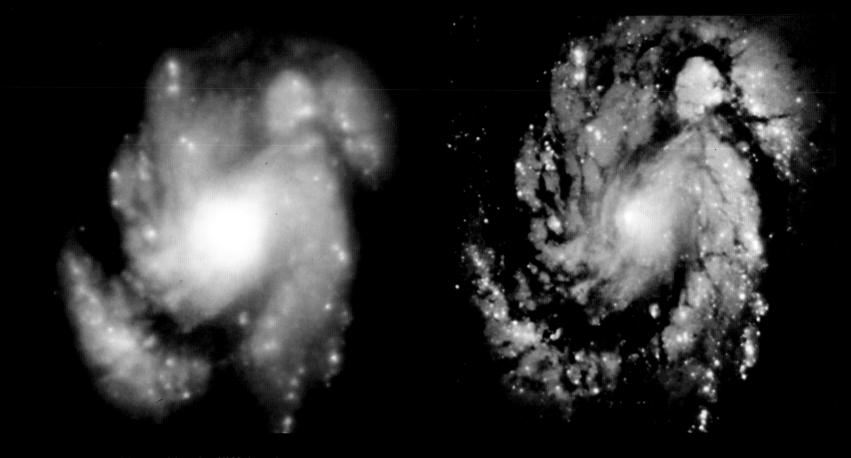

This comparison image of the core of the galaxy M100 shows the dramatic improvement in Hubble Space Telescope's view of the universe after the first Hubble Servicing Mission in December 1993. The new image, taken with the second generation Wide Field Planetary Camera (WFPC2) installed during the STS61 Hubble Servicing Mission, demonstrates that the camera's corrective optics compensate fully for the optical aberration in Hubble's primary mirror. With the new camera, the Hubble explored the universe with unprecedented clarity and sensitivity, and fulfilled its most important scientific objectives for which the telescope was originally built.

This is an image of a small portion of the Cygnus Loop supernova remnant, which marks the edge of a bubble-like, expanding blast wave from a colossal stellar explosion, occurring about 15,000 years ago. The Hubble image shows the structure behind the shock waves, allowing astronomers for the first time to directly compare the actual structure of the shock with theoretical model calculations. Besides supernova remnants, these shock models are important in understanding a wide range of astrophysical phenomena, from winds in newly-formed stars to cataclysmic stellar outbursts. The supernova blast is slamming into tenuous clouds of insterstellar gas. This collision heats and compresses the gas, causing it to glow, and the shock thus acts as a searchlight revealing the structure of the interstellar medium.

Glittering stars and wisps of gas create a breathtaking backdrop for the self-destruction of a massive star, called supernova 1987A, in the Large Magellanic Cloud, a nearby galaxy. Astronomers in the Southern hemisphere witnessed the brilliant explosion of this star on February 23, 1987. Shown in this NASA Hubble Space Telescope image, the supernova remnant, surrounded by inner and outer rings of material, is set in a mass of ethereal, diffuse clouds of gas. This three-colour image is composed of several pictures of the supernova and its neighbouring region taken with the Wide Field and Planetary Camera 2 in September 1994, February 1996 and July 1997. The numerous bright blue stars near the supernova are giant stars, each one over six times more massive than our Sun.

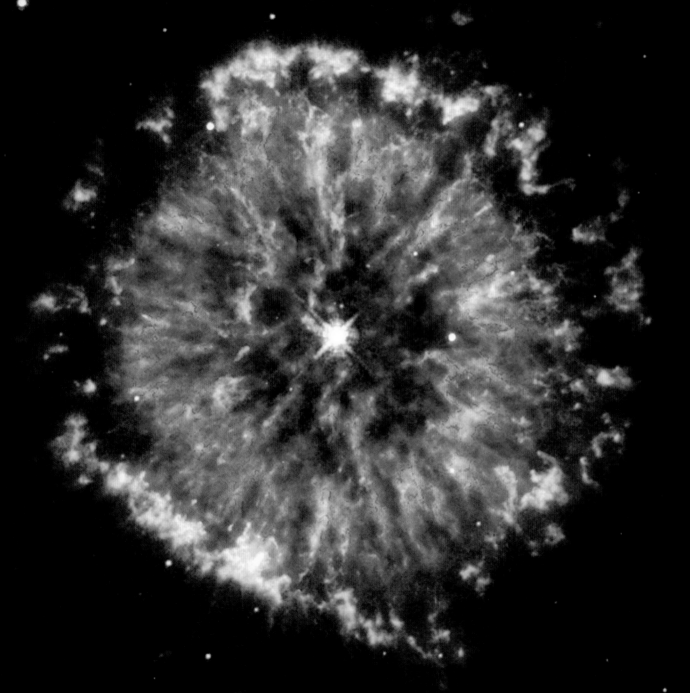

Glowing in the constellation Aquila like a giant celestial eye, planetary nebula NGC 6751 is a cloud of gas ejected several thousand years ago from the hot star visible in its centre. Planetary nebulae have nothing to do with planets. They are shells of gas thrown off by Sun-like stars nearing the ends of their lives. The star's loss of its outer, gaseous layers exposes the hot stellar core with its strong ultraviolet radiation. This causes the ejected gas to fluoresce as the planetary nebula.

BEYOND THE SOLAR SYSTEM

This eerie and dramatic picture from the Hubble telescope shows newborn stars emerging from 'eggs.' These are dense, compact pockets of interstellar gas called evaporating gaseous globules (EGGs). Hubble found the EGGs, appropriately enough, in the Eagle nebula, a nearby star-forming region 7,000 light-years from Earth in the constellation Serpens. These striking pictures, taken in 1995, show the EGGs at the tip of finger-like features protruding from monstrous columns of cold gas and dust in the Eagle Nebula (also called M16). The columns, dubbed 'elephant trunks,' protrude from the wall of a vast cloud of molecular hydrogen, like stalagmites rising above the floor of a cavern. Inside the gaseous towers, which are light-years long, the interstellar gas is dense enough to collapse under its own weight, forming young stars that continue to grow as they accumulate more and more mass from their surroundings.

The larger and more massive galaxy seen here is NGC 2207 (on the left
in the Hubble Heritage image), and the smaller one on the right is IC 2163.
Strong tidal forces from NGC 2207 have distorted the shape of IC 2163,
flinging out stars and gas into long streamers stretching out 100,000 light-
years toward the right-hand edge of the image. The high resolution of the
Hubble telescope image reveals dust lanes in the spiral arms of NGC 2207,
clearly silhouetted against IC 2163, which is in the background. Hubble also
reveals a series of parallel dust filaments extending like fine brush strokes
along the tidally stretched material on the right-hand side.

This spectacular colour panorama of the centre of the Orion nebula is one of the largest pictures ever assembled from individual images taken with the Hubble Space Telescope. The picture, seamlessly composited in 2002 from a mosaic of 15 separate fields, covers an area of sky about 5 percent of the area covered by the full Moon. The seemingly infinite tapestry of rich detail revealed by Hubble shows a churning turbulent star factory set within a maelstrom of flowing, luminescent gas. Although this 2.5 light-years wide view is still a small portion of the entire nebula, it includes almost all of the light from the bright glowing clouds of gas and a star cluster associated with the nebula.

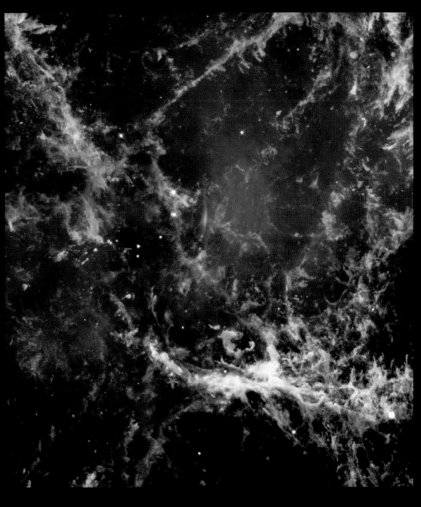

Penetrating 25,000 light-years of obscuring dust and myriad stars, Hubble provided the clearest view to date of one of the largest young clusters of stars inside the Milky Way galaxy, located less than 100 light-years from the very centre of the Galaxy. Having an equivalent mass greater than 10,000 stars like our sun, the monster cluster is ten times larger than typical young star clusters scattered throughout the Milky Way. It is destined to be ripped apart in just a few million years by gravitational tidal forces in the galaxy's core, but in its brief lifetime it shines more brightly than any other star cluster in the Galaxy. This image was taken in infrared light by Hubble's NICMOS camera in September 1997.

In the year 1054 AD, Chinese astronomers were startled by the appearance of a new star, so bright that it was visible in broad daylight for several weeks. Today, the Crab Nebula is visible at the site of the Guest Star. Located about 6,500 light-years from Earth, the Crab Nebula is the remnant of a star that began its life with about 10 times the mass of our own Sun. Its life ended on July 4, 1054 when it exploded as a supernova. In this image, NASA's Hubble Space Telescope has zoomed in on the centre of the Crab to reveal its structure with unprecedented detail. The Crab Nebula data were obtained by Hubble's Wide Field and Planetary Camera 2 in 1995. Images taken with five different colour filters have been combined to construct this new false-colour picture, which shows ragged shards of gas that are expanding away from the explosion site at over 3 million miles per hour.

This Hubble image shows one of the most complex planetary nebulae ever seen, NGC 6543, nicknamed the Cat's Eye Nebula. Hubble reveals surprisingly intricate structures including concentric gas shells, jets of high-speed gas and unusual shock-induced knots of gas. Estimated to be 1,000 years old, the nebula is a visual fossil record of the dynamics and late evolution of a dying star. This colour picture, taken with the Wide Field Planetary Camera 2, is a composite of three images taken at different wavelengths (red, hydrogen-alpha; blue, neutral oxygen, 6300 angstroms; green, ionized nitrogen, 6584 angstroms). The image was taken on September 18, 1994.

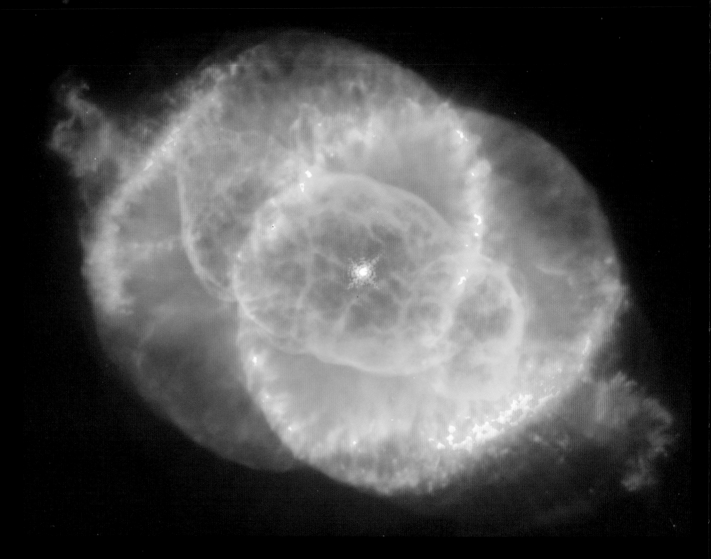

Hubble's spectacular resolution has allowed astronomers to resolve, for the first time, hot blue stars deep inside an elliptical galaxy. This swarm of nearly 8,000 blue stars resembles a blizzard of snowflakes near the core (lower right) of the neighbouring galaxy M32, located 2.5 million light-years away in the constellation Andromeda. Hubble confirms that the ultraviolet light comes from a population of extremely hot helium-burning stars at a late stage in their lives. Unlike the Sun, which burns hydrogen into helium, these old stars exhausted their central hydrogen long ago, and now burn helium into heavier elements. The observations, taken in October 1998, were made with the camera mode of the Space Telescope Imaging Spectrograph (STIS) in ultraviolet light.

BEYOND THE SOLAR SYSTEM |

This Hubble image reveals a pair of one-half light-year long interstellar 'twisters,' eerie funnels and twisted-rope structures in the heart of the Lagoon Nebula (Messier 8), which lies 5,000 light-years away in the direction of the constellation Sagittarius. The central hot star, O Herschel 36 (lower right), is the primary source of the ionizing radiation for the brightest region in the nebula, called the Hourglass. Analogous to the spectacular phenomena of Earth tornadoes, the large difference in temperature between the hot surface and cold interior of the clouds, combined with the pressure of starlight, may produce strong horizontal shear to twist the clouds into their tornado-like appearance.

In this stunning picture of the giant galactic nebula NGC 3603, Hubble captured various stages of the life cycle of stars in one single view. To the upper left of centre is the evolved blue supergiant called Sher 25. The star has a unique circumstellar ring of glowing gas that is a galactic twin to the famous ring around the supernova 1987A. This true-colour picture was taken on March 5, 1999 with the Wide Field Planetary Camera 2

One of the largest Hubble Space Telescope images ever made of a complete galaxy was unveiled by Hubble scientists in January 2005. The Hubble telescope captured a display of starlight, glowing gas, and silhouetted dark clouds of interstellar dust in this four-foot-by-eight-foot image of the barred spiral galaxy NGC 1300.

In this detailed view from Hubble, taken in September 2004, the so-called Cat's Eye Nebula, formally catalogued NGC 6543, lives up to its billing as one of the most complex such nebulae seen in space.

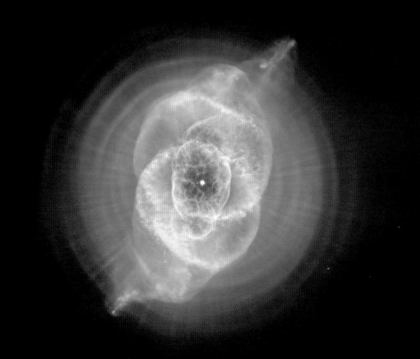

Taken in March 2004, this is the deepest portrait of the visible universe ever achieved by humankind. Called the Hubble Ultra Deep Field (HUDF), the million-second-long exposure reveals the first galaxies to emerge from the so-called 'dark ages,' the time shortly after the big bang when the first stars reheated the cold, dark universe. It's anticipated that the image will offer new insights into what types of objects reheated the universe long ago. This historic new view is actually two separate images taken by Hubble's Advanced Camera for Surveys (ACS) and the Near Infrared Camera and Multi-object Spectrometer (NICMOS). Both images reveal galaxies that are too faint to be seen by ground-based telescopes, or even in Hubble's previous faraway looks, called the Hubble Deep Fields (HDFs), taken in 1995 and 1998.

BEYOND THE SOLAR SYSTEM

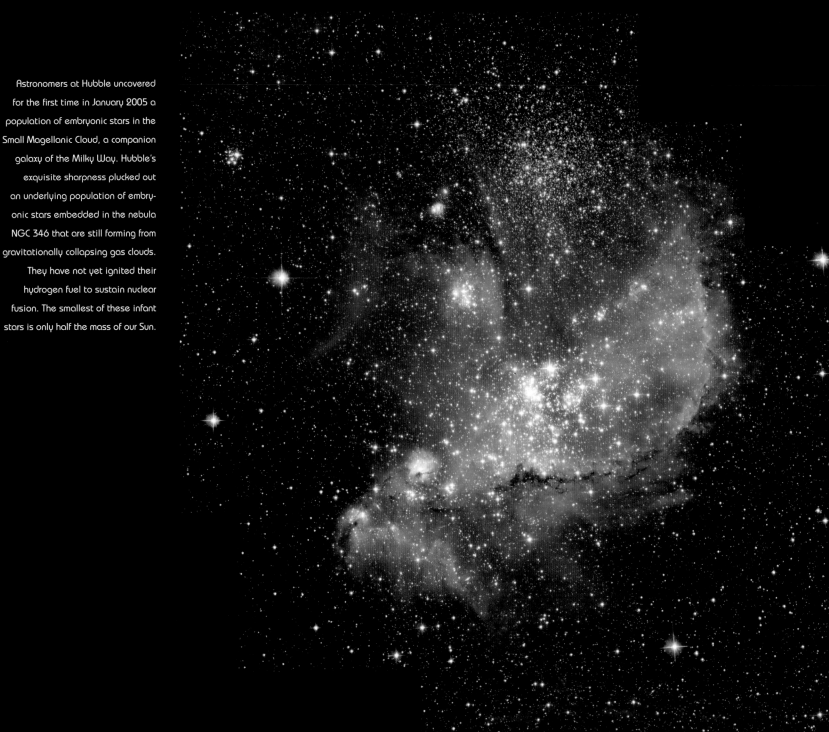

Astronomers at Hubble uncovered
for the first time in January 2005 a
population of embryonic stars in the
Small Magellanic Cloud, a companion
galaxy of the Milky Way. Hubble's
exquisite sharpness plucked out
an underlying population of embryonic stars embedded in the nebula
NGC 346 that are still forming from
gravitationally collapsing gas clouds.
They have not yet ignited their
hydrogen fuel to sustain nuclear
fusion. The smallest of these infant
stars is only half the mass of our Sun.

In this unusual image, NASA's Hubble Space Telescope captures a rare
view of the celestial equivalent of a geode – a gas cavity carved by the
stellar wind and intense ultraviolet radiation from a hot young star. Real
geodes are baseball-sized, hollow rocks that start out as bubbles in
volcanic or sedimentary rock. Only when these inconspicuous round rocks
are split in half by a geologist is there a chance to appreciate the inside
of the rock cavity that is lined with crystals. In the case of Hubble's
35 light-year diameter 'celestial geode' the transparency of its bubble-
like cavity of interstellar gas and dust reveals the treasures of its interior.

PART TWO: LOOKING BACK

One of the main benefits of Man's venture into space has been the ability to look back at the Earth, and to see our planet from a new perspective. This has allowed valuable data about such things as the climate, pollution levels and crop distribution to be gathered from an unassailable viewpoint, along with military information key to national defence. Like it or not, there is no hiding place from the eye in the sky, and the modern satellite is capable of resolving even the smallest detail with remarkable accuracy.

Satellites can take up fixed geostationary positions or form polar orbits. Some, such as Telstar and Intelsat, are there to act as an aid to communications, allowing data conversations to be relayed around the world. The function of others is to perform a variety of scientific missions, and typical examples of contemporary satellites are Terra and its sister craft Aqua. Their role is to explain such things as weather systems, leading to a greater understanding of why such phenomena as droughts and flooding occur, and allowing warnings to be given when dramatic storms are approaching.

One of the most important instruments that both Terra and Aqua carry is the Moderate-resolution Imaging Spectroradiometer (MODIS), which sees every point on the world every one to two days in 36 discrete spectral bands, and tracks a wide array

of the Earth's vital signs. MODIS can measure the photosynthetic activity of land and marine plants to yield better estimates of how much of the greenhouse gas is being absorbed and used in plant productivity, and it also maps the aerial extent of snow and ice brought by winter storms and frigid temperatures, and the 'green wave' that sweeps across continents as winter gives way to spring and vegetation blooms.

MODIS also sees where and when natural disasters – such as volcanic eruptions, floods, severe storms, droughts and wildfires – happen, and has the potential to help those on the ground get out of the way in time. Its bands are particularly sensitive to fires: they can distinguish flaming from smouldering burns and can provide better estimates of the amounts of aerosols and gases that fires release into the atmosphere. Meanwhile, the weather instruments aboard the Aqua satellite allow the most accurate, highest resolution measurements ever from space of the infrared brightness (radiance) of Earth's atmosphere. This data comes from two microwave sounding instruments that are part of the Atmospheric Infrared Sounder (AIRS) experiment: the Atmospheric Infrared Sounder itself and the Advanced Microwave Sounding Unit. With its visible, infrared and microwave detectors, AIRS provides a three-dimensional look at Earth's weather. Working in tandem, its instruments can make simultaneous observations for space all the way to Earth's surface, even in the presence of heavy clouds, and the system can create a global, three-dimensional map of atmospheric temperature and humidity. Through this, AIRS provides information about clouds, greenhouse gasses and many other atmospheric phenomena.

Not every picture from space is produced in such a highly technical way. In the introduction to this book we looked at the role of astronauts in producing photographs that were not only very often outstandingly beautiful, but which revealed information that was key to our understanding of Earth, and the relationship that all those who inhabit the planet have with each other. Since the Mercury missions of the early 1960s, astronauts have used hand held cameras to photograph the Earth and, to date, almost half a million exposures have been made in this way. The International Space Station (ISS) continues the tradition, with hand held photography starting from this location in November 2000. The ISS is well suited for observations: its average altitude is 220 miles above the Earth, and its orbital inclination of 51.6 degrees includes most of the coastlines and heavily populated areas of the world.

Surprising though it may seem, several of the images you'll see in this section were made by astronauts using cameras that, although professionally specified, are remarkably similar to those you would find in a photographer's studio on Earth. The Destiny US Laboratory Module has a science window with a clear aperture 50.8cm in diameter that is perpendicular to the Earth's surface most of the time. The window's three panes of fused silica are of optical quality, and this allows astronauts to achieve images that have enough quality and detail to satisfy the requirements of scientists back on Earth. Typical subjects would be major deltas in South and East Asia, coral reefs, major cities, smog over industrial regions, areas that typically experience floods or droughts triggered by El Nino cycles and alpine glaciers.

Hand-held photography fills a niche between aerial photography and imagery from satellite sensors, and can often provide valuable additional information. Dramatic events can be recorded with astronauts and scientists exchanging real-time information, while critical environmental regions will be photographed repeatedly over time, and can be compared with photographic records that date back to Gemini and Skylab missions.

GEOGRAPHY |

This perspective view shows Mount Ararat in easternmost Turkey, which has been the site of several searches for the remains of Noah's Ark. The main peak, known as Great Ararat, is the tallest peak in Turkey, rising to 5165 metres (16,945 feet). This image was generated from a Landsat satellite image draped over an elevation model produced by the Shuttle Radar Topography Mission (SRTM). The view uses a 1.25-times vertical exaggeration to enhance topographic expression, and the natural colours of the scene are enhanced by image processing, inclusion of some infrared reflectance (as green) to highlight the vegetation pattern, and inclusion of shading of the elevation model to further highlight the topographic features.

PREVIOUS PAGE
Clouds and sunglint (sun reflecting
off the water surface) over the
Indian Ocean as seen during the
STS-96 mission from the Space
Shuttle Discovery.

View of the Earth as seen by the Apollo 17 crew travelling towards the Moon. This translunar coast photograph extends from the Mediterranean Sea area to the Antarctica South polar ice cap. This is the first time the Apollo trajectory made it possible to photograph the South polar ice cap. Note the heavy cloud cover in the Southern Hemisphere, and the fact that almost the entire coastline of Africa is clearly visible. The Arabian Peninsula can be seen at the Northeastern edge of Africa. The large island off the coast of Africa is the Malagasy Republic, and the Asian mainland is on the horizon towards the Northeast.

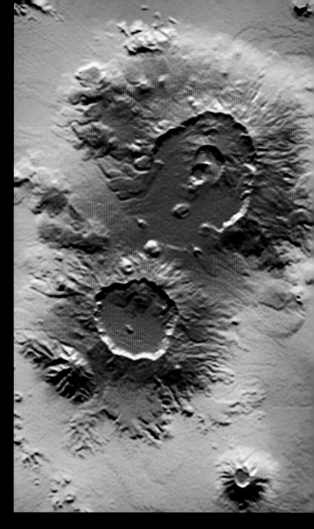

On February 19, 2000, Space Shuttle Endeavour passed over the highly active and dangerous volcanic zone of the Andes in Equador. Endeavour was mapping elevations on most of the Earth's land surface during the Shuttle Radar Topography Mission (SRTM). There have been more than 50 eruptions of Mount Cotopaxi alone since 1738. With its height of 5,897 metres (over 19,000 feet), it is more than 3,000 metres higher than the surroundings. The digital elevation model acquired by SRTM, with its resolution of 25 x 25 metres, is so rich in detail that you can even make out an inner crater with a diameter of 120 metres by 250 metres inside the outer crater (800 metres by 650 metres). Blue and green correspond to the lowest elevations in the image, while beige, orange, red, and white represent increasing elevations.

The area known as the Afar Triangle is located at the northern end of the East Africa Rift where it approaches the southeastern end of the Red Sea and the southwestern end of the Gulf of Aden. The East African Rift, the Red Sea and the Gulf of Aden are all zones where the Earth's crust is pulling apart in a process known as crustal spreading. Their three-way meeting is called a triple junction, and their spreading creates a triangular topographic depression for which the area was named. Shown here are a few of the volcanoes of the Afar Triangle. The larger two are Nabro Volcano (upper right, in Eritrea) and Mallahle Volcano (lower left, in Ethiopia). Elevation data used in this image was acquired by the Shuttle Radar Topography Mission (SRTM) aboard the Space Shuttle Endeavour

This perspective view shows the major volcanic group of Bali, one 13,000 islands comprising the nation of Indonesia. The conical mountain to the left is Gunung Agung, at 3,148 metres (10,308 feet) the highest point on Bali and an object of great significance in Balinese religion and culture. Agung underwent a major eruption in 1963 after more than 100 years of dormancy, resulting in the loss of over 1,000 lives. In the centre is the complex structure of Batur volcano, showing a caldera (volcanic crater) left over from a massive catastrophic eruption about 30,000 years ago. Two visualization methods were combined to produce the image: shading and colour coding of topographic height. Elevation data used in this image were acquired by the Shuttle Radar Topography Mission aboard the Space Shuttle Endeavour.

The STS-92 Space Shuttle astronauts photographed upstate New York at sunset on October 21, 2000. Water bodies (Lake Ontario, Lake Erie, the Finger Lakes, the St Lawrence and Niagara Rivers) are highlighted by sunglint (sun reflecting off the water surface), making for a dramatic and unusual regional view. The photograph was taken looking toward the southwest from southern Canada, and captures a regional smog layer extending across central New York, western Lake Erie and Ohio, and further west. The layer of atmospheric pollution is capped by an atmosphere inversion, which is marked by the layer of clouds at the top of the photograph. The astronauts were able to document this smog event from a variety of vantage points as they orbited over the north-eastern US and southern Canada.

A plume of steam or ash is still visible from Mount Etna in this SeaWiFS image captured on July 27, 2001. The plume is mingling with aerosols from other sources such as the Sahara and Europe. The somewhat clearer air over the Baltic Sea reveals the phytoplankton bloom that continues to fill much of the surface water.

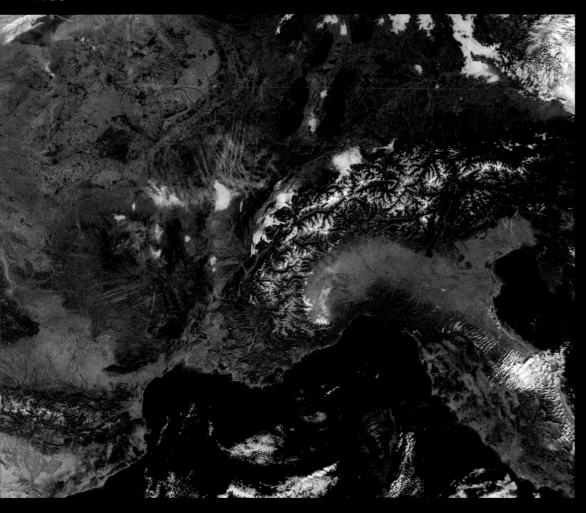

The MODIS instrument on board the Terra satellite collected this image of Western Europe in 2001. Starting at the lower-left corner of the image and moving clockwise are the countries of Spain, France, Germany, Austria, Italy, and Switzerland. The Alps, the crescent-shaped mountain range running from the centre to the right side of the image, span a length of 1,200 kilometres (750 miles) and cover an area of 128,750 square kilometres (80,000 square miles). In this image, a grey cloud of aerosols, predominantly from Italy's north-western region of Lombardy, are corralled by the massive heights of the Alps. One large source of aerosols is the city of Milan. Home to numerous international and local industrial plants, Milan is faced with many of the same air quality problems as other large metropolitan areas. Also visible in the image is a phytoplankton bloom in the Bay of Biscay (left), at the confluence of the Garonne and Dordogne rivers. The rivers form the Gironde Estuary, which is saturated with sediment that provides necessary nutrients

Bilbao, capital of Provincia de Vizcaya in northern Spain, consists of old and new sections connected by several bridges spanning the Rio Nervion. The old part of the city occupies the east bank of the river, and its modern offshoot, dating from the late nineteenth century, sits opposite. In 1983 a flood severely damaged the old section of Bilbao, but it has since been restored. This false-colour infrared composite image of the city was acquired by the ASTER instrument, flying aboard the Terra satellite in 2000, and the scene spans a total area 25 kilometres (15½ miles) wide by 24 kilometres (15 miles) tall. The reds and dark reds show vegetated areas, while the dark blue area (upper left) is where the waters of the Rio Nervion empty into the Bay of Biscay. Bright white areas are clouds, and it is possible to see their shadows on the surface slightly offset toward the north west. Light blue patches show the human developments in and around Bilbao.

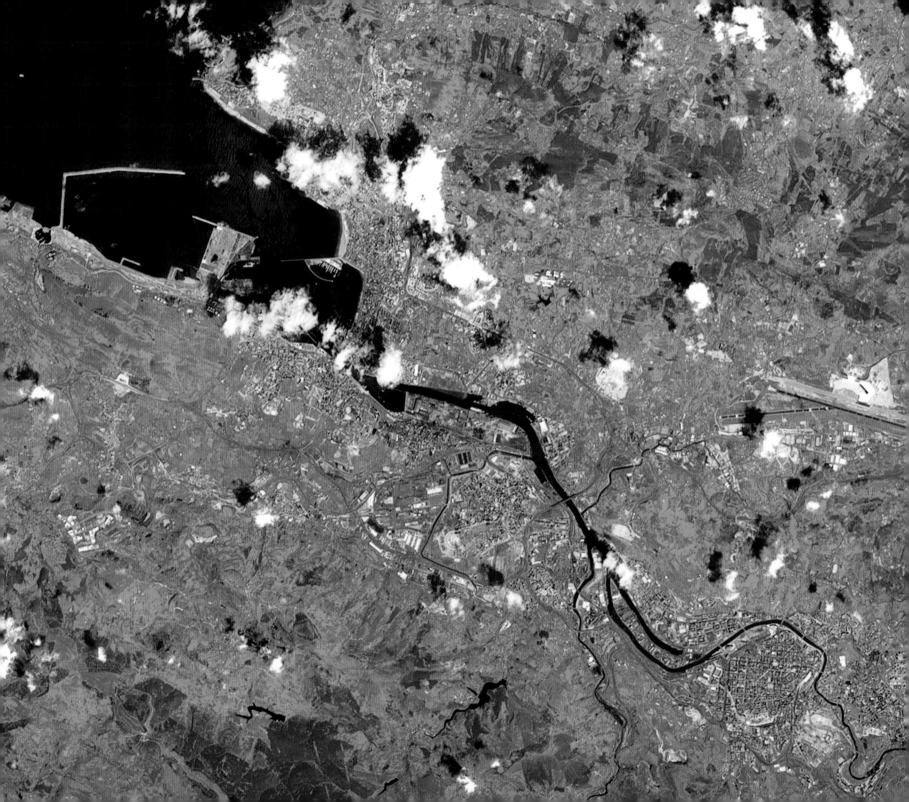

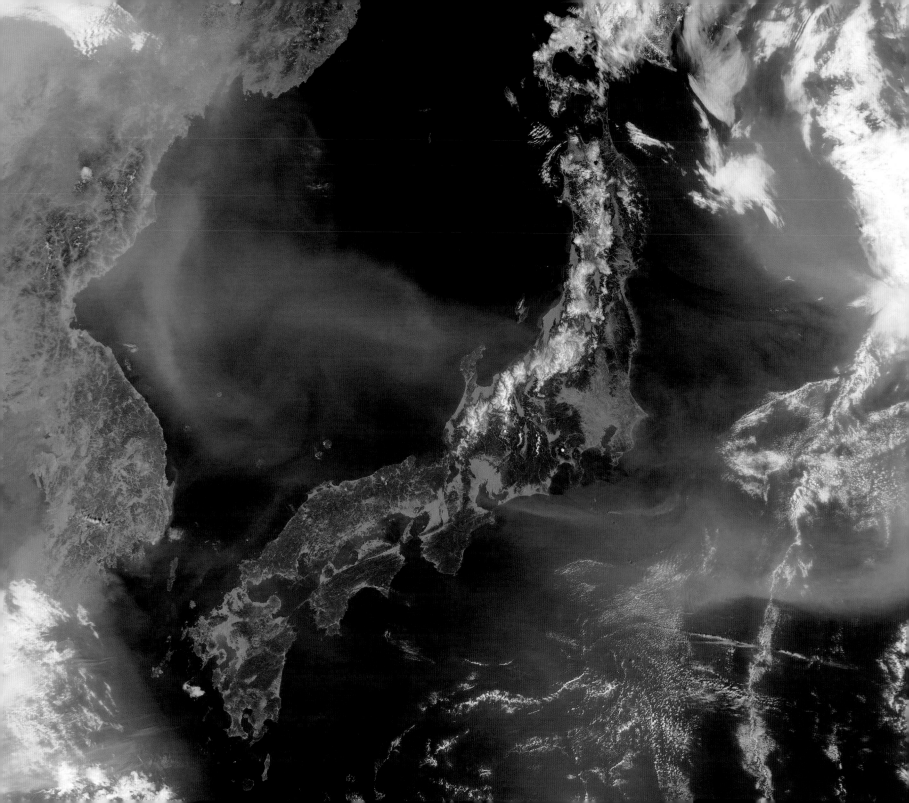

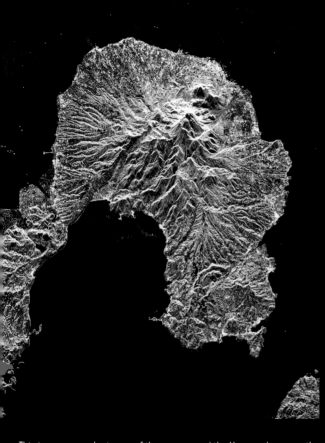

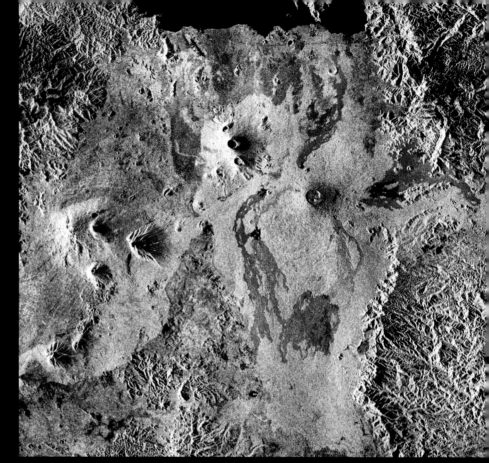

This is a space radar image of the area around the Unzen volcano, on the west coast of Kyushu Island in southwestern Japan. Unzen, which appears in this image as a large triangular peak with a white flank near the centre of the peninsula, has been continuously active since a series of powerful eruptions began in 1991. The image was acquired by the Spaceborne Imaging Radar-C/X-band Synthetic Aperture Radar (SIR-C/X- SAR) aboard the space shuttle Endeavour on its 93rd orbit on April 15, 1994.

Mount Oyama, on the Japanese island of Miyake-Jima, is erupting. Scientists estimate the volcano is emitting significant quantities of supher dioxide at a rate of between 10,000 and 20,000 tons per day. This true-colour scene was acquired during a small eruption on April 2, 2002, by the MODIS instrument aboard the Terra satellite. The last major eruption of Miyake-Jima occurred in

This is a false-colour composite of Central Africa, showing the Virunga volcano chain along the borders of Rwanda, Zaire and Uganda. The image was acquired on October 3 1994, on orbit 58 of the space shuttle Endeavour by the Spaceborne Imaging Radar-C/X band Synthetic Aperture Radar (SIR-C/X-SAR). The image covers an area 56 kilometres by 70 kilometres (35 miles by 43 miles). The dark area at the top of the image is Lake Kivu, which forms the border between Zaire (to the right) and Rwanda (to the left). In the centre of the image is the steep cone of Nyiragongo volcano, rising 3,465 metres (11,369 feet) high, with its central crater now occupied by a lava lake. To the left are three volcanoes, Mount Karisimbi, rising 4,500 metres (14,800 feet) high; Mount Sab inyo, rising 3,600 metres (12,000 feet) high; and Mount Muhavura, rising 4,100 metres (13,500 feet) high. To their right is Nyamuragira volcano, which is 3,053 metres (10,017

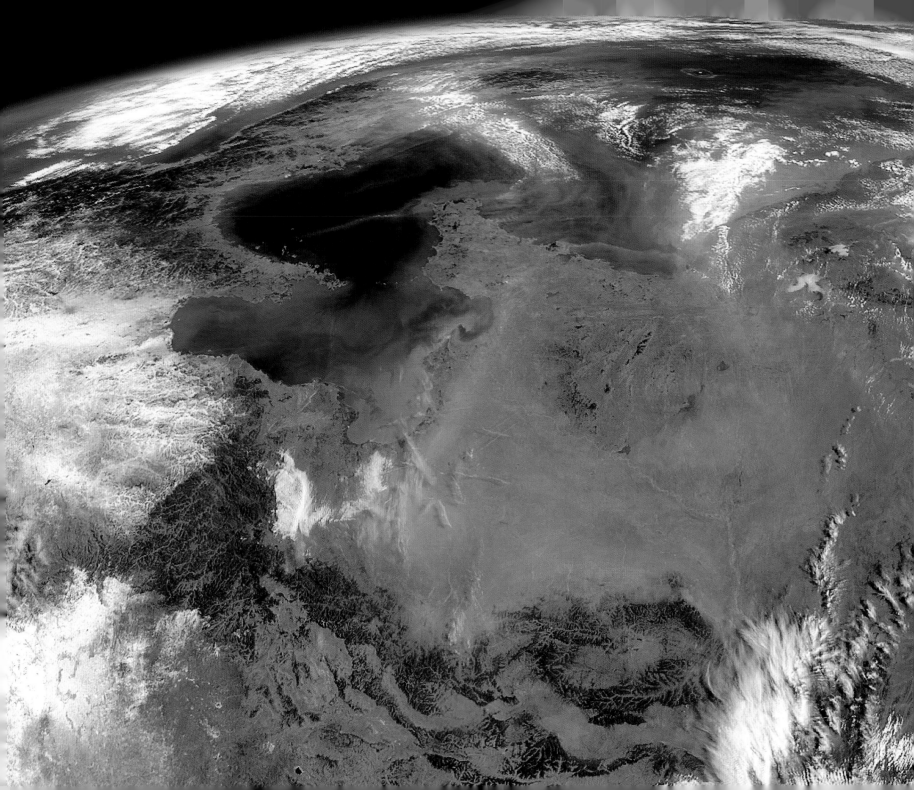

This 1990 SeaWifS image shows dense haze over eastern China. The view looks eastward across the Yellow Sea towards Korea.

This Advanced Spaceborne Thermal Emission and Reflection Radiometer (ASTER) sub-image, taken in 2001, covers a 12 x 12 kilometre (7½ x 7½ miles) area in Northern Shanxi Province, China. The low sun angle and light snow cover highlight a section of the Great Wall, visible as a black line running diagonally through the image from lower left to upper right. The wall is more than 2,000 years old and was built over a period of 1,000 years. Stretching 7,240 kilometres (4,500 miles) from Korea to the Gobi Desert, it was first built to protect China from marauders from the north.

This MODIS image from May 14, 2002, shows the marked contrasts between the high, arid Tibetan Plateau (upper left) and lush southern slopes of the Himalaya Mountains that give way to the Brahmaputra River Valley in northern India. MODIS detected several fires burning in the region, and they are marked with red dots.

The Himalayan Mountain Range runs a curving path from west to east in this true-colour Terra MODIS image from October 27, 2002. In this image, the Range separates southeastern China from India, and runs through (from left to right) northwestern India, Nepal, a small part of northeastern India, and Bhutan. In the top half of the image, a number of lakes glow like jewels scattered throughout southeastern China's Plateau of Tibet. Many of the lakes show tinges of blue-green that probably indicate microscopic plant life. In the bottom half of the image, a number of major rivers flow to the southeast, eventually joining the Ganges and emptying into the Bay of Bengal (not shown). Meanwhile, in northern India, a grey haze of air pollution hangs over some of the most densely populated cities in the world.

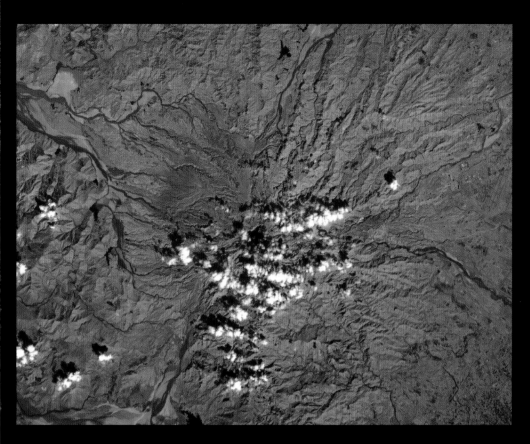

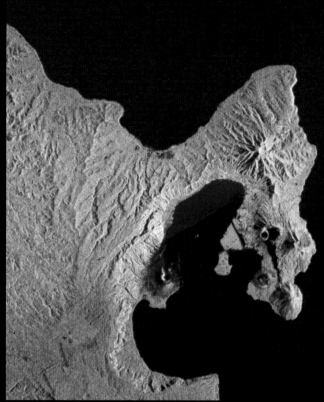

On June 15, 1991 Mount Pinatubo in the Philippines exploded in the most violent volcanic eruption of the twentieth century. Over the following decade, scientists used NASA remote sensing technology to study both the effects of the eruption on climate and the way the volcano and its surrounding area changed over time. This Landsat 7 false-colour image of Mount Pinatubo, acquired in January 2001, was made using a combination of shortwave infrared, near infrared, and green light. In the centre of the image (partially obscured by clouds) is Pinatubo Crater Lake, which was formed by rainwater collecting in the caldera left by the 1991 eruptions. Heat from the volcano warms the water to 24°C (75°F).

This is a radar image of the Rabaul volcano on the island of New Britain, Papua New Guinea, taken almost a month after its September 19, 1994, eruption that killed five people and covered the town of Rabaul and nearby villages with up to 75 centimetres (30 inches) of ash. More than 53,000 people were displaced by the eruption. The image was acquired by the Spaceborne Imaging Radar-C/X-band Synthetic Aperture Radar (SIR-C/X-S AR) aboard the space shuttle Endeavour on its 173rd orbit on October 11, 1994.

GEOGRAPHY

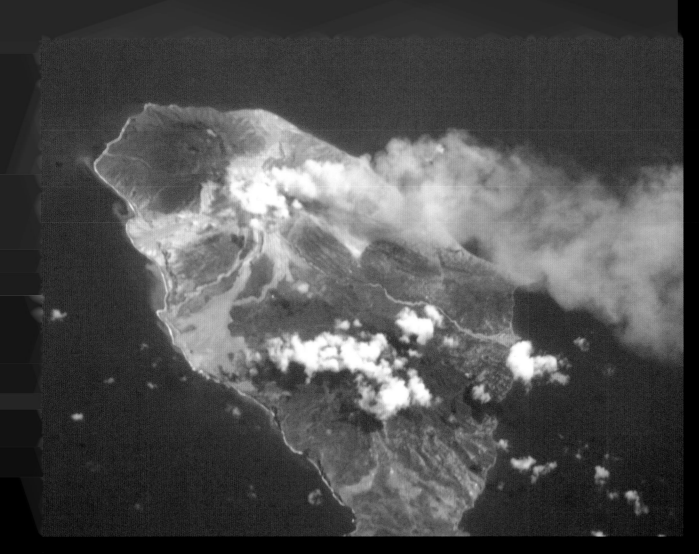

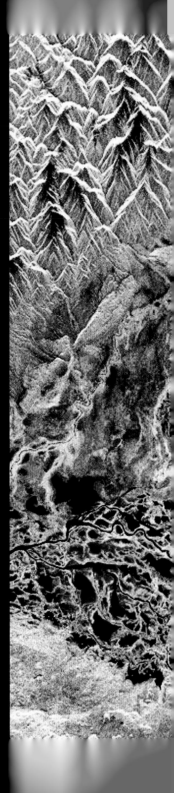

Volcanic activity on the West Indian island of Montserrat has remained high for several years – the current activity started in 1995. However, remote sensing of the island has been difficult because of frequent cloud cover. The International Space Station crew flew north of the island on a clear day in early July 2001 and recorded a vigorous steam plume emanating from the summit of Soufriere Hills. The image also reveals how rivers draining the mountain have brought down large amounts of volcanic debris and caused extensive volcanic mud flows (lahars) and new deltas to be built out from the coast. On a small island like Montserrat which is only 13 kilometres by 8 kilometres (8 miles by 5

This is an image of the area of Kliuchevskoi volcano, Kamchatka, Russia, which began to erupt on September 30, 1994. Kliuchevskoi is the blue triangular peak in the centre of the image, towards the left edge of the bright red area that delineates bare snow cover. The image was acquired by the Spaceborne Imaging Radar-C/X-band Synthetic Aperture Radar (SIR-C/X-SAR) aboard the space shuttle Endeavour on its 88th orbit on October 5, 1994, and the image shows an area approximately 75 kilometres by

Waves of clouds along the east flanks of the Andes Mountains cast off an orange glow, created by the low angle of the sun in the west. This view was photographed in 1965 by astronauts Frank Borman and James A Lovell during the Gemini VII mission, looking south from Northern Bolivia across the Andes. The Intermontane Salt Basins are visible in the background.

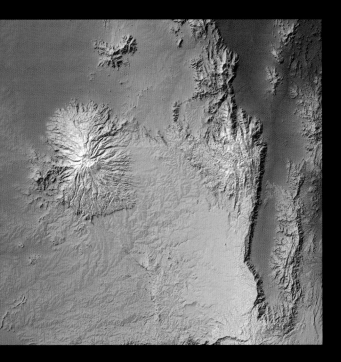

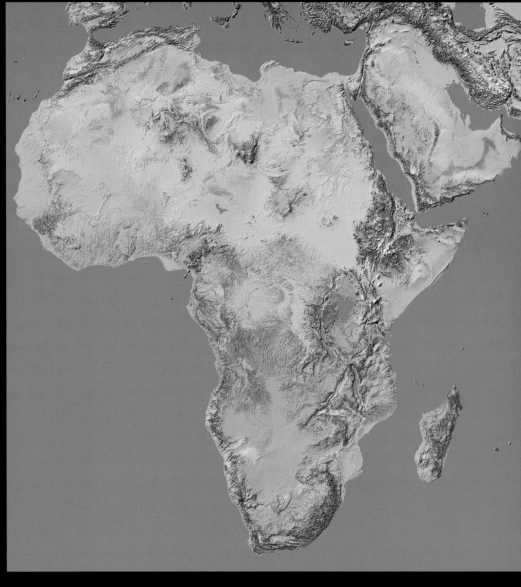

The striking contrast of geological structures in Africa is shown in this shaded relief image of Mount Elgon on the left and a section of the Great Rift Valley on the right. Mount Elgon is a solitary extinct volcano straddling the border between Uganda and Kenya, and at 4,321 metres (14,178 feet) tall is the eighth highest mountain in Africa. Juxtaposed with this impressive mountain is a section of the Great Rift Valley, a geological fault system that extends for about 4,830 kilometres (2,995 miles) from Syria to central Mozambique. Two visualization methods, shading and colour coding of topographical height, were combined to produce the image, and the elevation data used here was acquired by the Shuttle Radar Topography Mission aboard the Space Shuttle Endeavour.

This colour shaded relief image shows the extent of digital elevation data for Africa captured by the Shuttle Radar Topography Mission (SRTM), which was designed to collect three-dimensional measurements of the Earth's surface. The instrument flew on board the Space Shuttle Endeavour in February 2000 and used an interferometric radar system to map the topography of Earth's landmass. This kind of digital elevation data is particularly useful to scientists studying earthquakes, volcanism, and erosion patterns, and for use in mapping and modelling hazards to human habitation.

Iceland's icy exterior hides its steamy volcanic underpinnings. Running roughly northeast to southwest through the island country is the northern part of the vast Mid-Atlantic Ridge, the divergent boundary of the North American tectonic plate and the Eurasian plate. The two plates are diverging, essentially pulling Iceland apart, and often resulting in intense sub-glacial volcanic activity. Beneath the enormous Vatnajvkull Glacier the intense geothermal activity continuously melts the ice to form a sub-glacial lake in the volcano's caldera (crater). The melt waters eventually flood the caldera, and pour out from underneath the glacier, resulting in often devastating glacial outburst floods that occur every 5–15 years. This MODIS image was made from data acquired on January 27, 2002.

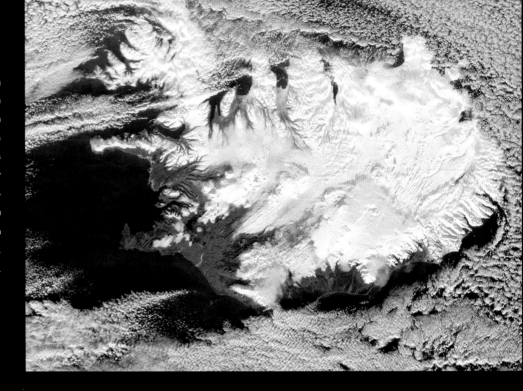

This set of images from the Multi-angle Imaging SpectroRadiometer shows the coastal states of Louisiana, Mississippi, Alabama and part of the Florida panhandle in 2001. The two smaller images, far right, are (top) a natural colour view comprised of red, green, and blue band data from MISR's nadir (vertical-viewing) camera, and (below) a false-colour view comprised of near-infrared, red, and blue band data from the same camera. The predominantly red colour of the false-colour image is due to the presence of vegetation, which is bright at near-infrared wavelengths. Cities appear as grey patches, with New Orleans visible at the southern edge of Lake Pontchartrain, along the left-hand side of the images. The large image is similar to the true-colour nadir view, except that red band data from the 60-degree backward-looking camera has been substituted into the red channel; the blue and green data from the nadir camera have been preserved.

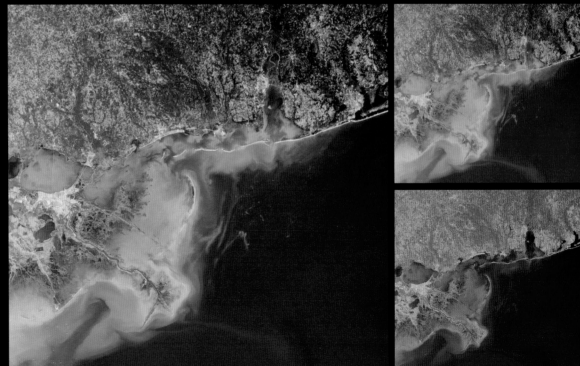

Home to more than four million people from over 100 countries, Sydney is the largest and most cosmopolitan of Australia's cities. This simulated true-colour Advanced Spaceborne Thermal Emission and Reflection Radiometer (ASTER) image shows the Sydney metropolitan area on October 12, 2001, and covers an area of 42 kilometres by 33 kilometres (26 miles by 20 miles). The image displays the concentrated development of the urban area with the harbour in the

upper right part of the image and the airport runways clearly visible protruding into Botany Bay to the south. Sustainable use of water and effective water management are prime concerns for the area. The city has developed numerous dams and water storage areas, including the Prospect Reservoir, located at the upper left edge of the image. Built in 1888, this reservoir is one of Sydney's oldest and was once the major water storage facility for the city.

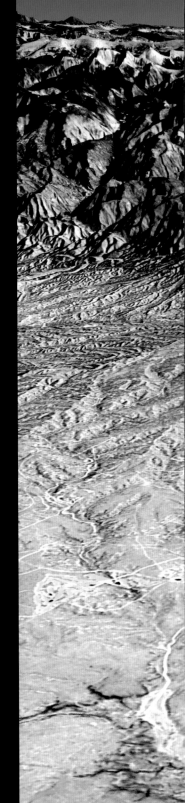

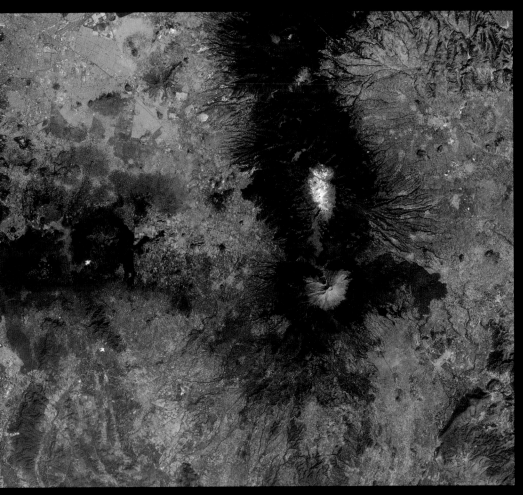

Located about 65 kilometres (40 miles) southeast of Mexico City, Popocatepetl roared back to life on December 18, 2000, spewing red hot rocks, ash, and smoke high into the air over the Valley of Mexico. Concerned that there may be even more massive eruptions to follow, or perhaps mudslides from the summit's melting snow and ice, Mexican authorities asked nearby residents to evacuate the region. This true-colour image of Popocatepetl was acquired on January 4, 1999, by the Enhanced Thematic Mapper Plus (ETM+) aboard NASA's Landsat 7 satellite. Even from this two-dimensional perspective, you get a sense of the volcano's impressive slopes as it towers some 5,465 metres (17,930 feet) above the surrounding landscape. Snow and ice encircle the summit, at the top of which the volcano's crater can be seen clearly.

Santa Fe, New Mexico, sits nestled in the foothills of the southern end of the Rocky Mountains at 2,134 metres (7,000 feet) above sea level. In this image, the Sangre de Cristo range of mountains appears in the background, and the city is seen from a western viewpoint. The data for this view was obtained from a collection of cloud-free scenes over all of North America collected by the Landsat 7 satellite's Enhanced Thematic Mapper Plus (ETM+) instrument in 2000, and these were stitched together by EarthSat Corporation for NASA. The subsection of the mosaic around Santa Fe was draped over a digital terrain model to give the appearance of a landscape.

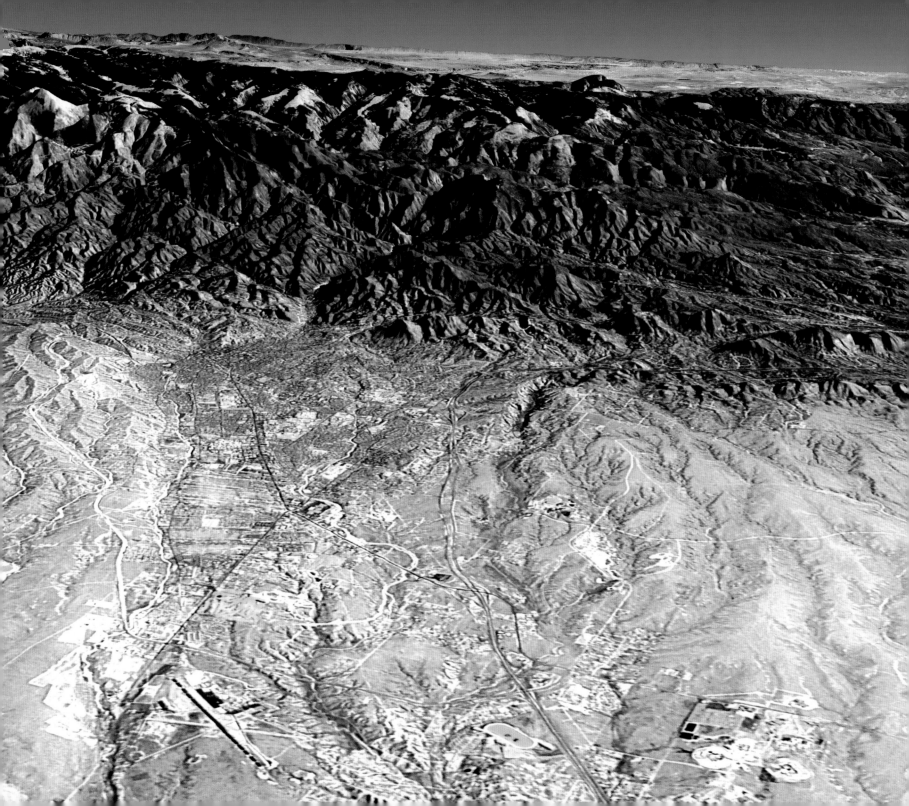

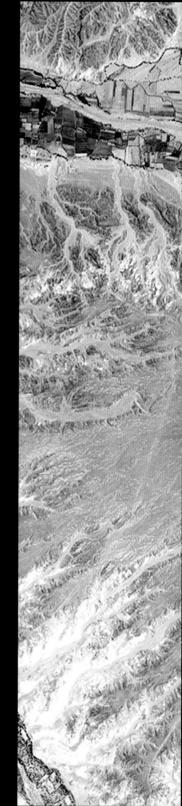

The Nasca Lines, recorded here by the ASTER instrument aboard NASA's Terra satellite in 2000, are located in the Pampa region of Peru, the desolate plain of the Peruvian coast 400 kilometres south of Lima. ASTER used visible and infrared spectral bands to create a simulated true colour image. The Lines were first spotted when commercial airlines began flying across the Peruvian desert in the 1920s, and passengers reported seeing 'primitive landing strips' on the ground below. The Lines were made by removing the iron-oxide coated pebbles which cover the surface of the desert. When the gravel is removed, they contrast with the light colour underneath, and in this way the lines were drawn as furrows of a lighter colour. On the Pampa, south of the Nasca Lines, archaeologists have now uncovered the lost city of the line-builders, Cahuachi. It was built nearly 2,000 years ago and mysteriously abandoned 500 years later.

This spaceborne radar image shows archeological sites, the environment that gave rise to them, and modern developments that threaten them in the region around Petra, Jordan, which is a World Heritage Site. The bright line across the centre of the image (running right to left or north to south) is a geological boundary between the limestone highlands (purple area) of Jebal Shara (the biblical Mount Sier) and deeply eroded sandstone steppes (green and orange area). For thousands of years, springs that occur along this boundary have provided water to the occupants of what are now some of the world's most significant archeological sites. The bluish green area above the line, near the centre of the image, marks the core area of the ancient caravan trading city of Petra, constructed in the sandstone canyons by the Nabataeans. Near the beginning of the Christian Era, this area controlled the spice and incense trade through the Arabian Peninsula for hundreds of years. The modern town of Wadi Musa is the bright orange and blue area below the bright line. This image was acquired by Spaceborne Imaging Radar-C/X-Band Synthetic Aperture Radar (SIR-C/X-SAR) on board the space shuttle Endeavour on April 9, 1994.

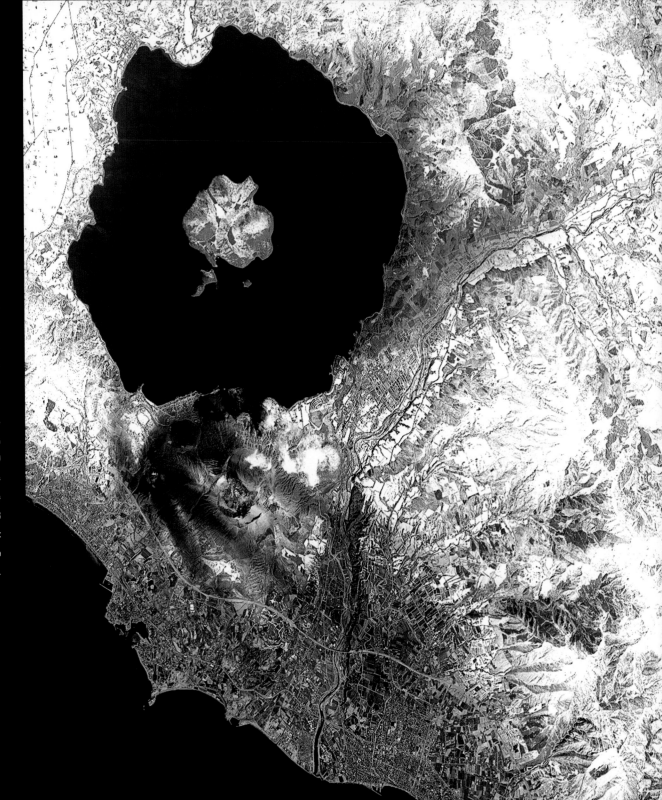

On April 3 2000, the Advanced Spaceborne Thermal Emission and Reflection Radiometer (ASTER) on the Terra Satellite captured this image of the erupting Mount Usu volcano in Hokkaido, Japan. On March 31 more than 11,000 people had been evacuated by helicopter, truck and boat from the foot of Usu, which had begun erupting from the northwest flank, shooting debris and plumes of smoke streaked with blue lightning thousands of feet into the air. Although no lava gushed from the mountain, rocks and ash continued to fall after the eruption. ASTER is the only high resolution imaging sensor on Terra. The primary goal of its mission is to obtain high-resolution image data in 14 channels over the entire land surface, as well as black and white stereo images. With a revisit time of between 4 and 16 days, ASTER provides the capability for repeat coverage of changing areas on Earth's surface.

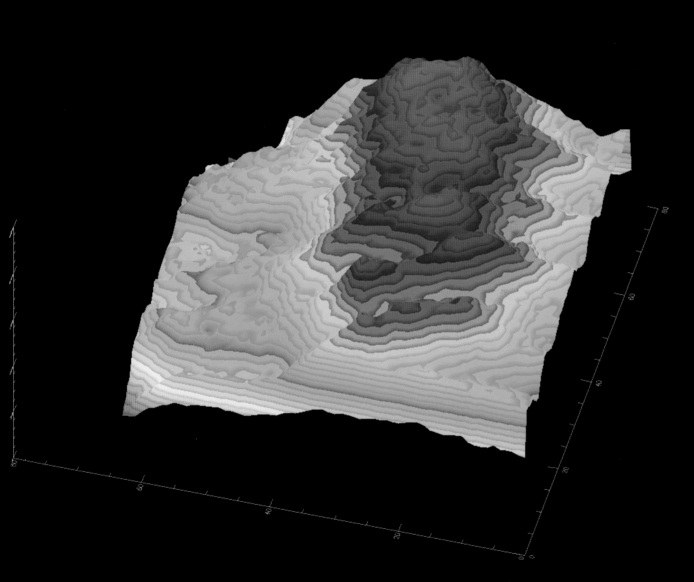

This three-dimensional image of the volcano Kilauea was generated based on interferometric fringes derived from two X-band Synthetic Aperture Radar data takes on April 13, 1994 and October 4, 1994. The altitude lines are based on quantitative interpolation of the topographic fringes. The level difference between neighbouring altitude lines is 20 metres (66 feet), and the ground area covers 12 kilometres by 4 kilometres (7½ miles by 2½ miles). The altitude difference in the image is about 500 metres (1,640 feet).

During the Golden Age of Greece the poet Homer wrote of the epic battles between Agamemnon and Priam – the fabled Trojan War. Although the city of Troy was a tourist attraction in Greek and Roman times, by the 1800s its location was lost, and many believed the story was only a myth. Frank Calvert and Heinrich Schliemann thought otherwise, and in the 1870s began excavating an earthen mound in western Turkey, near the Dardanelles. The site did indeed turn out to be the legendary city of Troy. Although the ruins of Troy have been explored for approaching 150 years, archaeologists have only excavated 10 percent of the site. To help them, NASA scientists are exploring new ways of using remote sensing data, and the image here shows Troy and the surrounding image in true colour. Taken by the Advanced Land Imager (ALI) aboard the EO-1 satellite, the full-size image has a resolution of 10 metres. These and other sensors may help find the boundaries of a harbour near Trojan-war era Troy that has since filled with sediment, trace the route of a Roman aqueduct that carried water to the city 2000 years ago, locate an ancient cemetery, and map the outer walls.

PART TWO: LOOKING BACK

This 1994 radar image shows the east coast of central Florida, including the Cape Canaveral area. The Indian River, Banana River and the Atlantic Ocean are the three bodies of water (shown in deep blue) from the lower left to the upper right of this false colour image. Parts of NASA's John F Kennedy Space Center (KSC) and the Cape Canaveral Air Station (CCAS) are visible. KSC occupies much of Merritt Island in the centre of the image, as well as the northern part of Cape Canaveral along the right side of the image. The light blue areas on Cape Canaveral are the launch pads used by NASA and the Air Force. The two pads in the upper left of the image (light blue hexagons with bright yellow areas in the middle) are Launch Complex 39 pads A and B, originally designed for the Apollo program and now used by the space shuttle. The other launch pads that dot the coastline are part of the CCAS and are used to launch robotic spacecraft, like the Cassini mission to Saturn. Two runways also appear as dark lines in the image. The one in the upper left is part of the space shuttle landing facility: at 4,572 metres (15,000 feet) long and 91.4 metres (300 feet) wide, it's one of the longest runways in the world.

This ASTER image of Mount Vesuvius, Italy was acquired on September 26, 2000, and the full size, false-colour image covers an area 36 kilometres by 45 kilometres (22½ miles by 28 miles). Vesuvius overlooks the city of Naples and the Bay of Naples in central Italy. (Popocatepetl and Mount Fuji are two other volcanoes surrounded by dense urban areas.) In 79 AD, Vesuvius erupted cataclysmically, burying all of the surrounding cities with up to 30 metres (98½ feet) of ash. The towns of Pompeii and Herculanaeum were rediscovered in the eighteenth century, and excavated in the twentieth century.

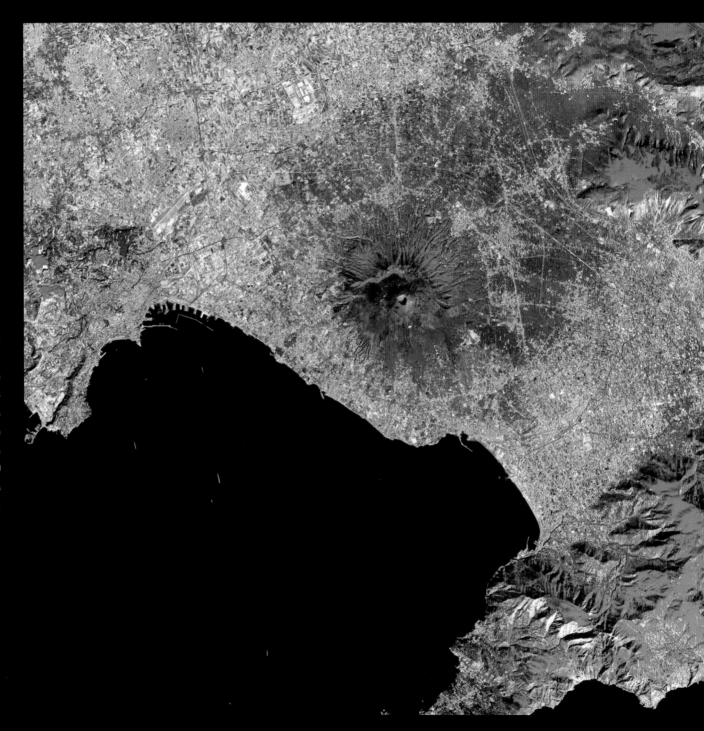

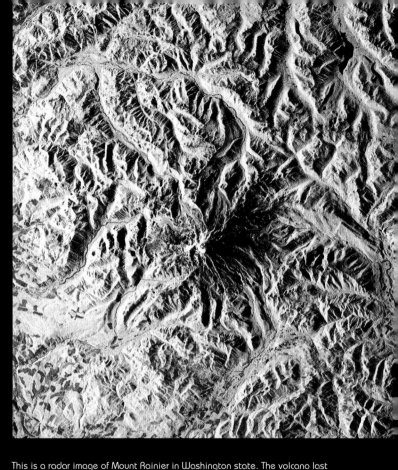

The catastrophic eruption of Mount St Helens on May 18, 1980 ranks among the most important natural events of the twentieth century in the United States. Because Mount St Helens is in a remote area of the Cascades Mountains, only a few people were killed by the eruption, but property damage and destruction totalled billions of dollars. The height of Mount St Helens was reduced from about 2950 metres (9677 feet) to about 2550 metres (8364 feet) as a result of the explosive eruption, which sent a column of dust and ash upwards more than 25 kilometres (15½ miles) into the atmosphere, and shock waves from the blast knocked down almost every tree within 10 kilometres (over 6 miles) of the central crater. The area of almost total destruction was about 600 square kilometres (373 square miles). This image was acquired by Landsat 7 almost exactly 20 years after the eruption, on August 22, 1999. It was produced at 30-metre resolution using bands 3, 2, and 1 to display red, green, & blue, respectively ('true colour'). Some of the effects of the massive eruption on May 18, 1980 can still be seen, although the rejuvenation process is obvious. Although very different to its former self, Mount St Helens is actively recovering

This is a radar image of Mount Rainier in Washington state. The volcano last erupted about 150 years ago and numerous large floods and debris flows have originated on its slopes during the last century. Today the volcano is heavily mantled with glaciers and snowfields, and more than 100,000 people live on young volcanic mudflows less than 10,000 years old and, consequently, are within the range of future, devastating mudslides. This image was acquired by the Spaceborne Imaging Radar-C and X-band Synthetic Aperture Radar (SIR-C/X-SAR) aboard the space shuttle Endeavour on its 20th orbit on October 1, 1994. The area shown in the image is approximately 59 kilometres by 60 kilometres (36½ miles by 37 miles). The scene was illuminated by the shuttle's radar from the northwest so that northwest-facing slopes are brighter and southeast-facing slopes are dark. Forested regions are pale green in colour; clear cuts and bare ground are bluish or purple; ice is dark green and white.

This spaceborne radar image shows the 'Valley Island' of Maui, Hawaii in 1994. The cloud-penetrating capabilities of radar provide a rare view of many parts of the island, since the higher elevations are frequently shielded from view by the atmosphere. The three major population centres, Lahaina on the left at the western tip of island, Wailuku left of centre, and Kihei in the lower centre appear as small yellow, white or purple mottled areas. West Maui volcano, in the lower left, is 1,800 metres high (5,900 feet) and is considered extinct. The entire eastern half of the island consists of East Maui volcano, which rises to an elevation of 3,200 metres (10,500 feet) and features a spectacular crater called Haleakala at its summit. The multi-wavelength capability of the SIR-C radar carried on board the space shuttle Endeavour also permitted differences in the vegetation cover on the middle flanks of East Maui to be identified. Rainforests appear in yellow, while grassland is shown in dark green, pink and blue.

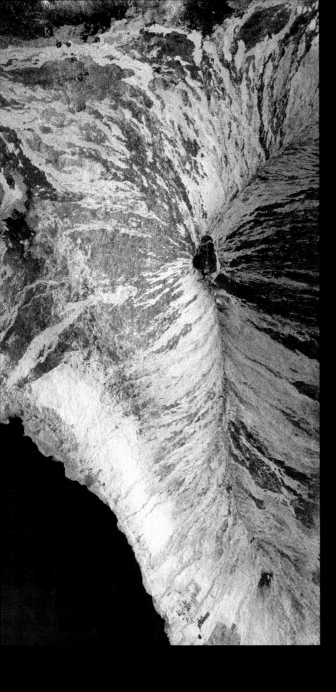

This image of the Mauna Loa volcano on the Big Island of Hawaii shows the capability of imaging radar to map lava flows and other volcanic structures. Mauna Loa has erupted more than 35 times since the island was first visited by westerners in the early 1800s. The large summit crater, called Mokuaweoweo Caldera, is clearly visible near the centre of the image. If the height of the volcano was measured from its base on the ocean floor instead of from sea level, Mauna Loa would be the tallest mountain on Earth. Its peak (centre of the image) rises over 8 kilometres (5 miles) above the ocean floor. This image was acquired by the Spaceborne Imaging Radar-C/X-Band Synthetic Aperture Radar (SIR-C/ X-SAR) aboard the space shuttle Endeavour on its 36th orbit on October 2, 1994.

Taken during the 1998 STS-95 mission from a point over Cuba, this photo shows an oblique, fore-shortened view of the Florida Peninsula, with the light blue, shallow seafloor of both the Florida Keys (curving across the bottom of the view) and the Bahama banks (right). 'Popcorn' cumulus cloud covers Miami and the Southern Everglades, although the built-up area from Fort Lauderdale to West Palm Beach can be discerned. Lake Okeechobee is the prominent water body in Florida, while Cape Canaveral can be seen half way up the peninsula. Orlando appears as the lighter patch west (left) of Cape Canaveral, near the middle of the peninsula. Cape Hatteras appears top right, with the north part of Chesapeake Bay also visible.

GEOGRAPHY

This image, taken on August 25, 1992 by the NOAA GOES-7 weather satellite, shows Hurricane Andrew making landfall on the Louisiana coast.

This view of the Earth shows a unique but physically impossible view of the southern hemisphere and Antarctica. While a spacecraft could find itself directly over the Earth's pole, roughly half of the image would be in darkness: this view was created by creating a mosaic from several images taken by the Galileo satellite over a 24 hour period and projecting them as they would be seen from above the pole. The continents of South America, Africa and Australia are respectively seen at the middle left, upper right, and lower right. The slightly bluish ice and snow of Antarctica includes large ice shelves (upper left, lower middle), a broad fan of broken offshore pack ice (lower left and middle) and continental glaciers protruding into the sea (lower right). The regularly spaced weather systems are prominent. Most spacecraft travelling near the Earth's poles are in very low Earth orbit, and cannot acquire panoramic shots like this one.

This picture of Houston, Texas illustrates the new detail being obtained for cities around the world by crew members on the International Space Station. This image, captured on December 17, 2000, centres on the downtown region and shows extensive detail of streets, parks and major buildings. The retractable roof of the new Astros baseball stadium, Enron Field, is open. Photography of cities to monitor urban growth is one of the objectives of NASA's Crew Earth Observations payload from the International Space Station.

Portland, the largest city in Oregon, is located on the
Columbia River at the northern end of the Willamette
Valley and, on clear days, Mount Hood highlights the
Cascade Mountains backdrop to the east. The Columbia
is the largest river in the American Northwest and is
navigable up to, and well beyond, Portland. It is also
the only river to fully cross the Cascade Range, and has
carved the Columbia River Gorge (seen towards the
left of the image). This perspective view of the region

46° Forward

26° Forward

Nadir

26° Backward

Depending upon the position of the Sun, the solar power stations in California's Mohave Desert can reflect solar energy from their large, mirror-like surfaces directly toward one of the Multi-angle Imaging SpectroRadiometer (MISR) cameras on board the Terra satellite. This allows them to appear dramatically brighter at some observation angles than at others. The solar power fields are readily discernible in this set of natural-colour images as the Sun's rays are reflected differently from the solar power fields at different observation angles. When these four images were acquired in 2003, the MISR camera closest to the specular reflection angle (the angle at which a perfect mirror reflects light) was the 26 forward-pointing camera. The solar fields can be readily identified by comparing the 26 forward camera view (top right-hand panel) with the other camera views. The two Solar Electric Generating Systems (SEGS) that appear alternately dim and very bright are the 150 megawatt array at Kramer Junction (slightly above image center) and the 160 megawatt array at Harper Lake (upper right-hand corner).

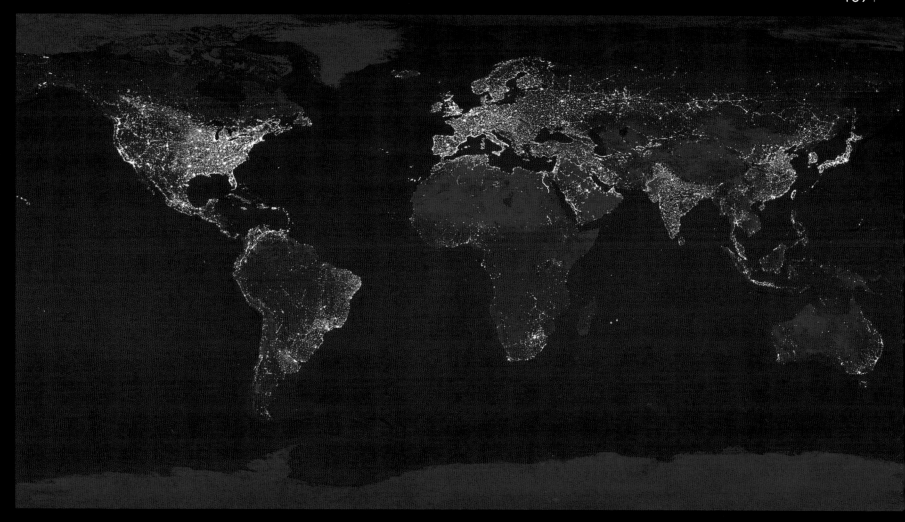

Growth in 'mega-cities' is altering the landscape and the atmosphere in such a way as to curtail normal photosynthesis. By using data from The Defence Meteorological Satellite Programme's Operational Linescan System, researchers have been able to look at urban sprawl by monitoring the emission of light from cities at night. By overlaying these 'light maps' onto other data such as soil and vegetation maps, the research shows that urbanisation can have a variable but measurable impact on photosynthetic productivity.

A composite image of the lights of the Earth's cities seen from space, with the Eastern United States, Europe, and Japan being brightly lit, while the interiors of Africa, Asia, Australia, and South America remain (for now) dark and lightly populated.

GEOGRAPHY |

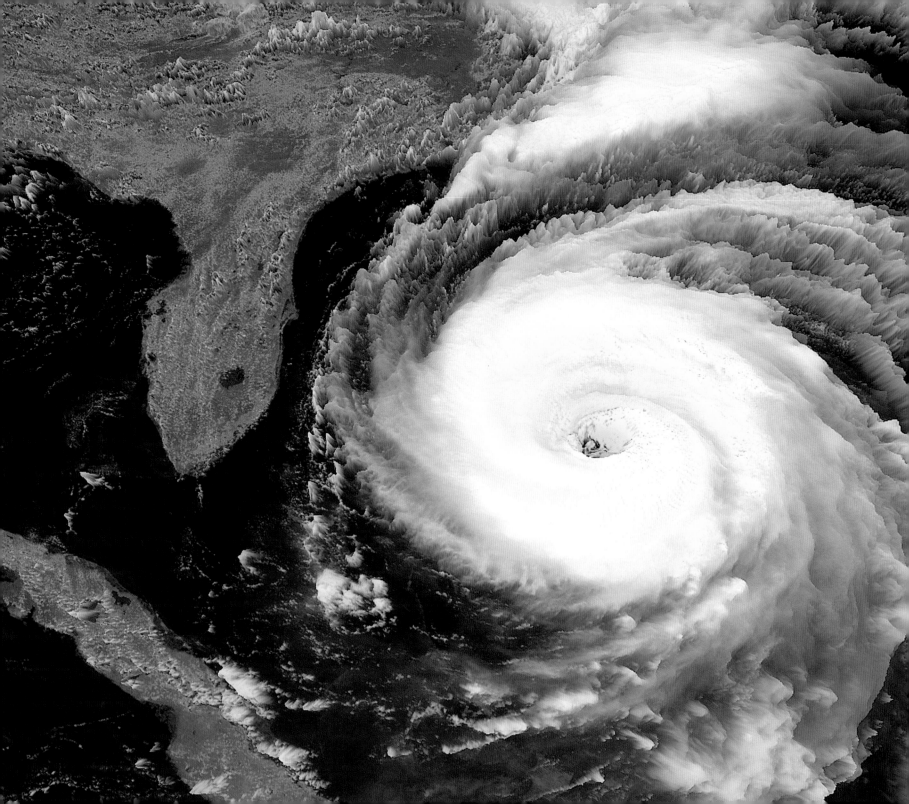

WEATHER AND CLIMATE

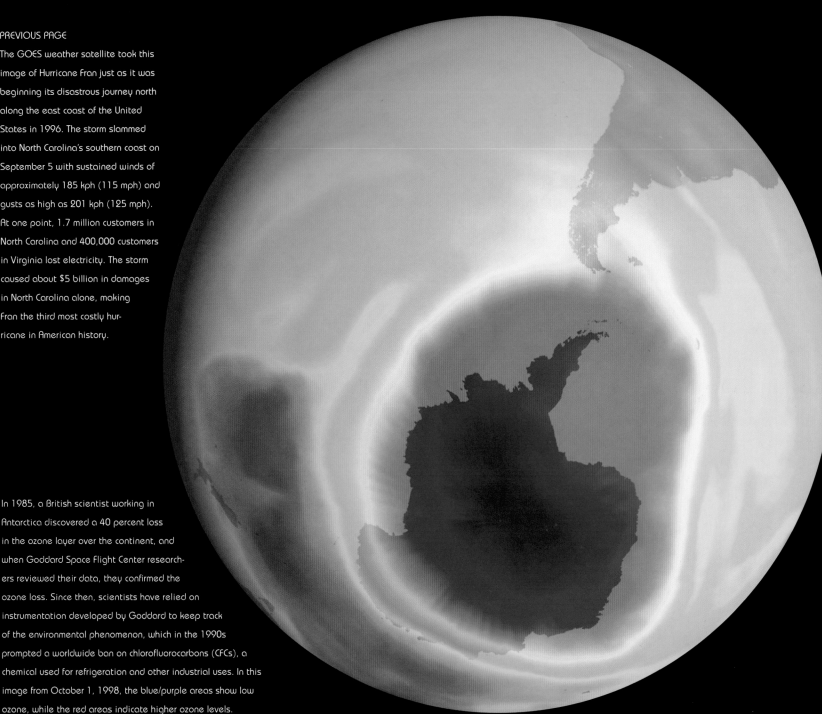

PREVIOUS PAGE
The GOES weather satellite took this image of Hurricane Fran just as it was beginning its disastrous journey north along the east coast of the United States in 1996. The storm slammed into North Carolina's southern coast on September 5 with sustained winds of approximately 185 kph (115 mph) and gusts as high as 201 kph (125 mph). At one point, 1.7 million customers in North Carolina and 400,000 customers in Virginia lost electricity. The storm caused about $5 billion in damages in North Carolina alone, making Fran the third most costly hurricane in American history.

In 1985, a British scientist working in Antarctica discovered a 40 percent loss in the ozone layer over the continent, and when Goddard Space Flight Center researchers reviewed their data, they confirmed the ozone loss. Since then, scientists have relied on instrumentation developed by Goddard to keep track of the environmental phenomenon, which in the 1990s prompted a worldwide ban on chlorofluorocarbons (CFCs), a chemical used for refrigeration and other industrial uses. In this image from October 1, 1998, the blue/purple areas show low ozone, while the red areas indicate higher ozone levels.

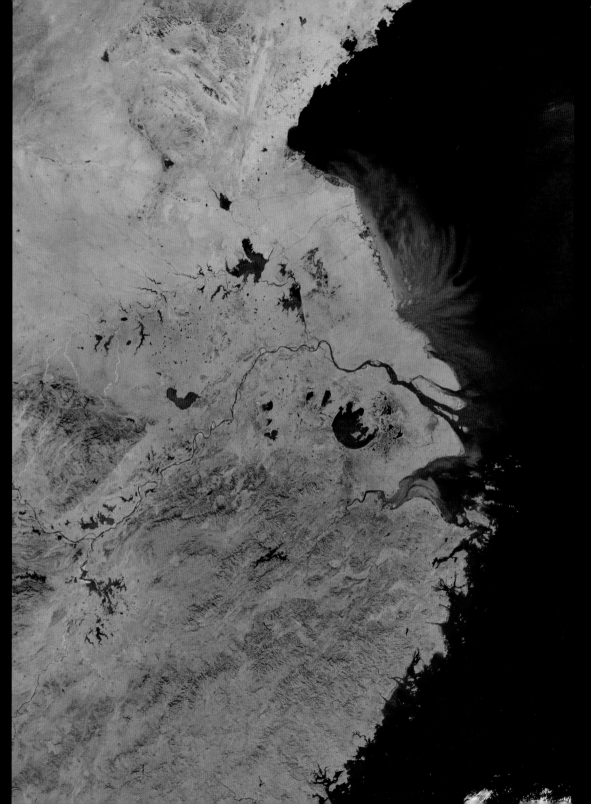

This combined true- and false-colour image from the MODIS instrument on the Aqua satellite shows thick air pollution (true-colour) and flooding (false-colour) in eastern China in 1993. The muddy flood waters blend in with the fading vegetation of the winter landscape and are further obscured by air pollution in the true-colour image, but the false-colour picks out the standing water which appears bright blue or black against the green vegetation.

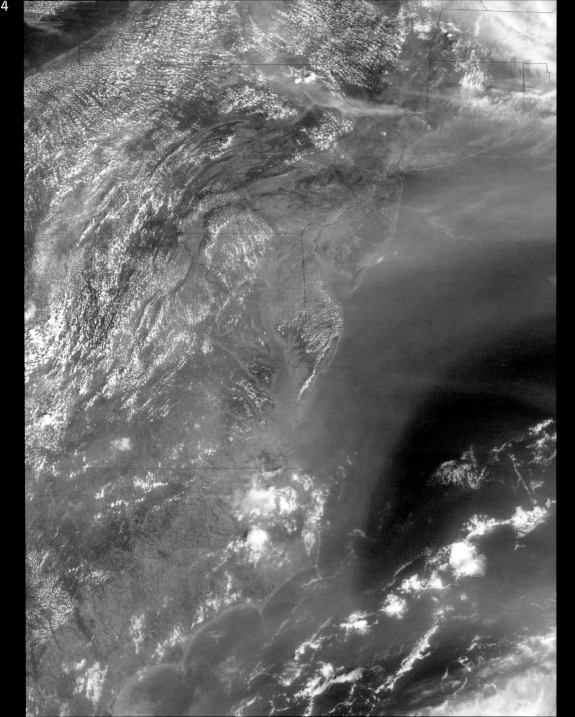

Air quality across the Great Lakes and Mid-Atlantic regions of the United States was less than ideal when this image was taken by the MODIS instrument in late June 2002. Much of the pollution is likely to be caused by smoke from forest fires in the west and fires in Canada's prairie provinces to the north. Since the dominant weather patterns across the country move air from west to east, it is not uncommon for air quality along the America's east coast to be affected by fires out west.

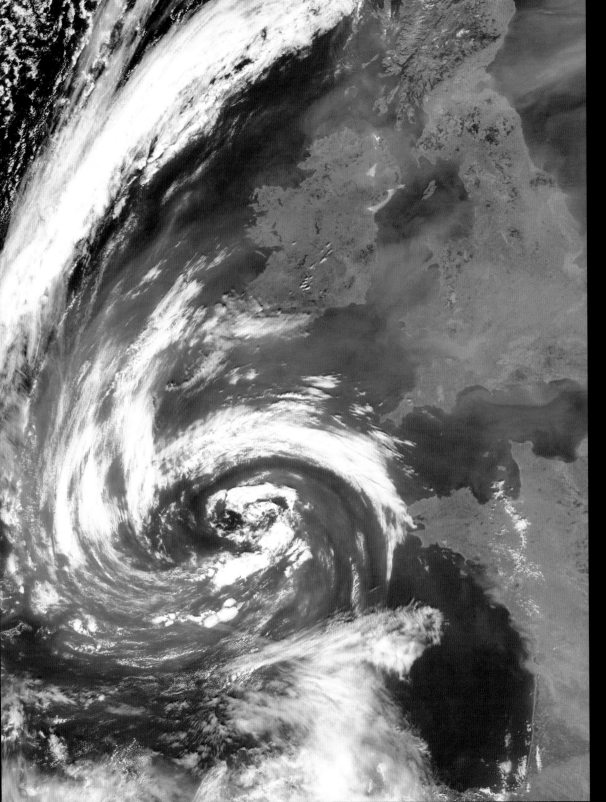

Pollution and smoke get caught up in a swirl of clouds off the coasts of Spain, France, Ireland, and the United Kingdom in this true-colour Aqua MODIS image acquired on March 23, 2003. The smoke and pollution appear as a greyish haze concentrated mostly over England and Ireland, blurring the landscape underneath. Despite the haze lying over the region, cloudy green and tan swirls of sediment and microscopic marine life are visible in the waters of the Atlantic, the English Channel, and in the Celtic (south of Ireland), Irish (east of Ireland), and North (east of England) Seas.

Air pollution is a severe and persistent problem in the foothills of the Himalaya Mountains in Pakistan, India, and Bangladesh. The haze and pollution back up against the mountains and remain for weeks at a time, posing a severe health hazard. In addition, scientists are beginning to gather evidence that the widespread and persistent nature of the pollution is even modifying the regional weather, particularly rainfall patterns. The pollution comes from inefficient wood and dung-fuelled heating and cooking devices, as well as forest fires and industrial and urban pollution.

WEATHER AND CLIMATE |

Korea and the Sea of Japan are obscured by swirls of pollution in this image taken by the Sea-viewing Wide Field-of-view Sensor (SeaWifS) on November 23, 2001.

A large crack is forming in the ice of the East Siberian Sea in this true-colour MODIS image from June 1, 2001. The thinnest ice appears bright blue because some light is able to penetrate and reflect off the water beneath.

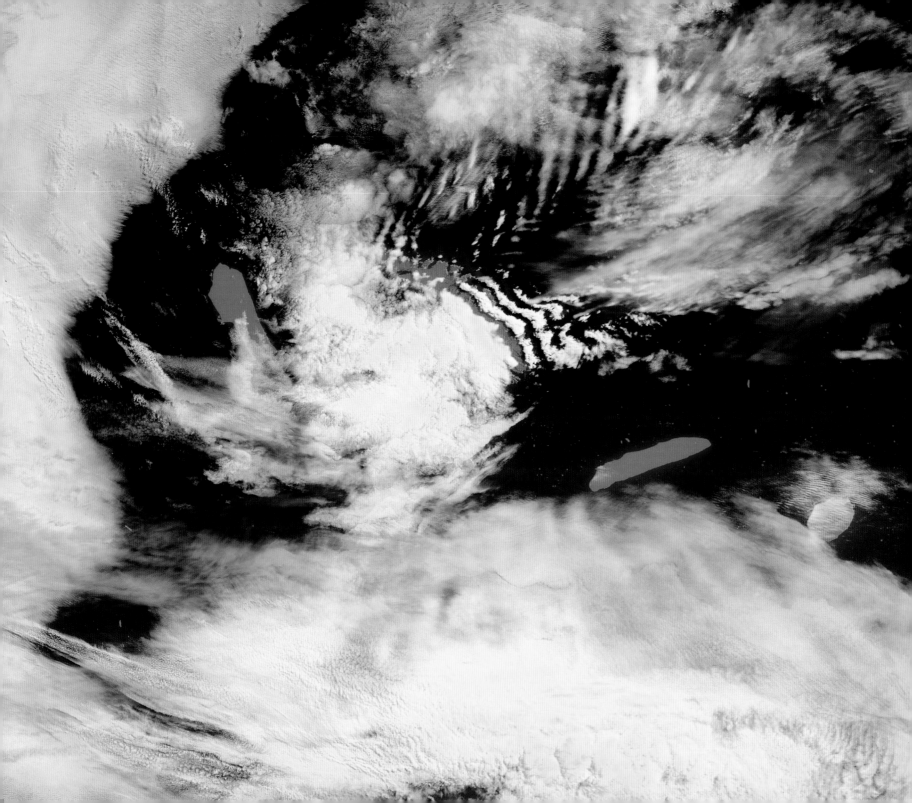

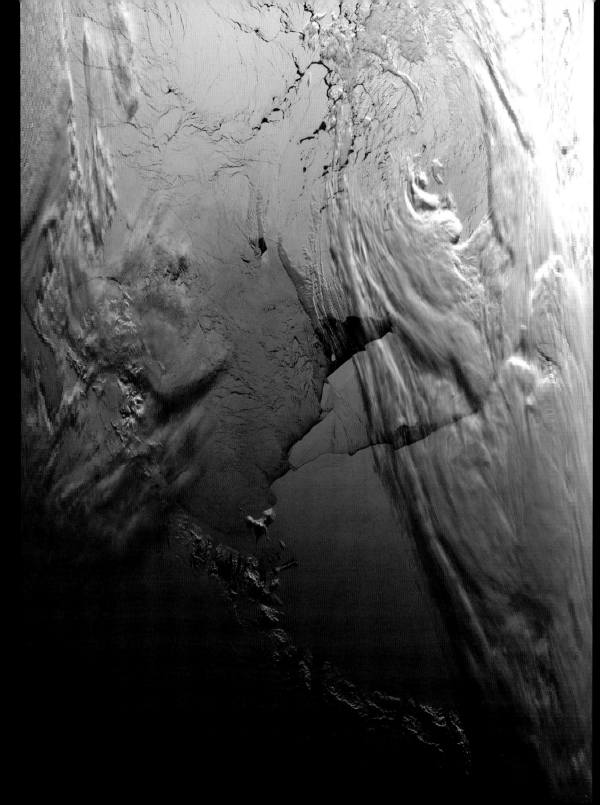

This false colour image from the MODIS instrument on board the Terra satellite shows the movement of a few large icebergs near South Georgia, an island off the east of the southern tip of South America. Here, ice id red, liquid water on the ground is dark blue to black and liquid water clouds are white. Despite its seemingly icy veneer, the mountainous island is an important wildlife habitat, harbouring seals, penguins and albatrosses.

Almost an iceberg 'nursery', icebergs continue to break away from the Ross Ice Shelf in Antarctica. This image from the MODIS instrument aboard the Terra spacecraft shows the level of activity along the shelf near Ross Island on September 21, 2000. The B-15 fragments are remnants of the huge iceberg, nearly 6,840 square kilometres (4,250 square miles), which broke away from the Antarctic shelf in late March 2000. Cracks in the Antarctic ice shelf are closely observed by satellite and are of interest to scientists studying the potential effects of global warming. This true-colour image was produced using MODIS bands 1, 3, and 4.

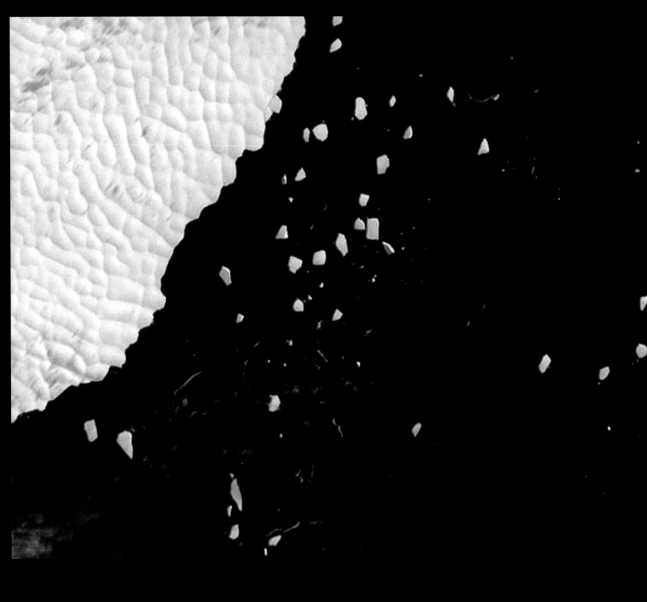

In the 1980s an iceberg over 1,600 kilometres (1,000 square miles) in area, and 400 metres (¼ mile) thick, broke off an Antarctic glacier and drifted into the Southern Ocean. The National Iceberg Center, which monitors sea ice in shipping lanes, named the giant B10. In 1995, B10 split in two. The larger piece (B10A) was still the size of Rhode Island. By mid-1999 it had moved out of the isolated waters around Antarctica and near the Drake Passage, used by ships to navigate around the southern tip of South America. This true-colour Landsat 7 image shows relatively small icebergs 'calving' off the edge of B10A. Since it is now in relatively warm water, it is breaking up more rapidly. The new icebergs are drifting dangerously into international shipping lanes. Remote sensing satellites such as Landsat 7, SeaWinds, and Radarsat are being used to monitor B10A and its child icebergs.

Warmer surface temperatures over just a few months in the Antarctic can splinter an ice shelf and prime it for a major collapse. This true-colour image from Landsat 7, acquired on February 21, 2000, shows pools of melt water on the surface of the Larsen Ice Shelf, and drifting icebergs that have split from the shelf.

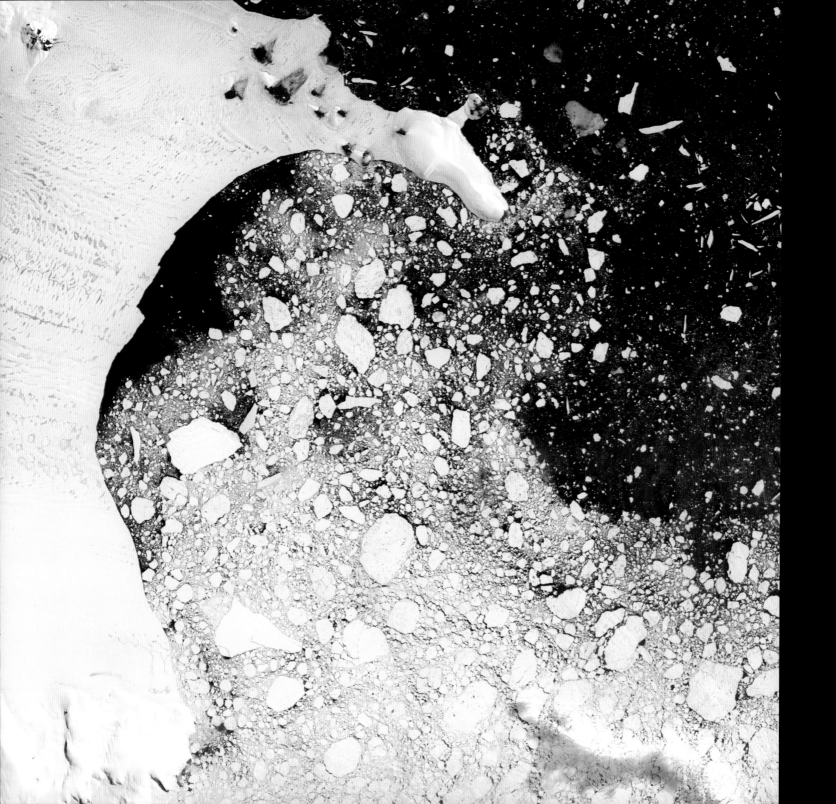

Clouds represent an area of great uncertainty in studies of global climate, and scientists are interested in better understanding the processes by which they form and change over time. In recent years, scientists have turned their attention to the ways in which human-produced aerosol pollution modifies clouds. One area that has drawn particular attention is 'ship tracks,' or clouds that form from the sulphate aerosols released by large ships. Although ships are not significant sources of pollution themselves, they do release enough sulphur dioxide in the exhaust from their smokestacks to modify overlying clouds. This image was acquired over the northern Pacific Ocean by the MODIS instrument aboard the Terra satellite in 2002.

This image taken by the MODIS instrument in late summer 2002 shows the opening up of the Davis Strait between western Greenland and Baffin Island, Canada. The Labrador Current, which flows southward into the Labrador Sea, has packed sea ice and icebergs up against the Baffin Island Coast (left). Snow cover is making its brief, summer retreat from the west coast of Greenland (right), exposing the rocky landscape. In the upper right of the image a low pressure centre is causing a whirlpool effect in the cloud bank against the Greenland coast. At the time this picture was taken the Strait was clear, and snow was retreating from the coastline.

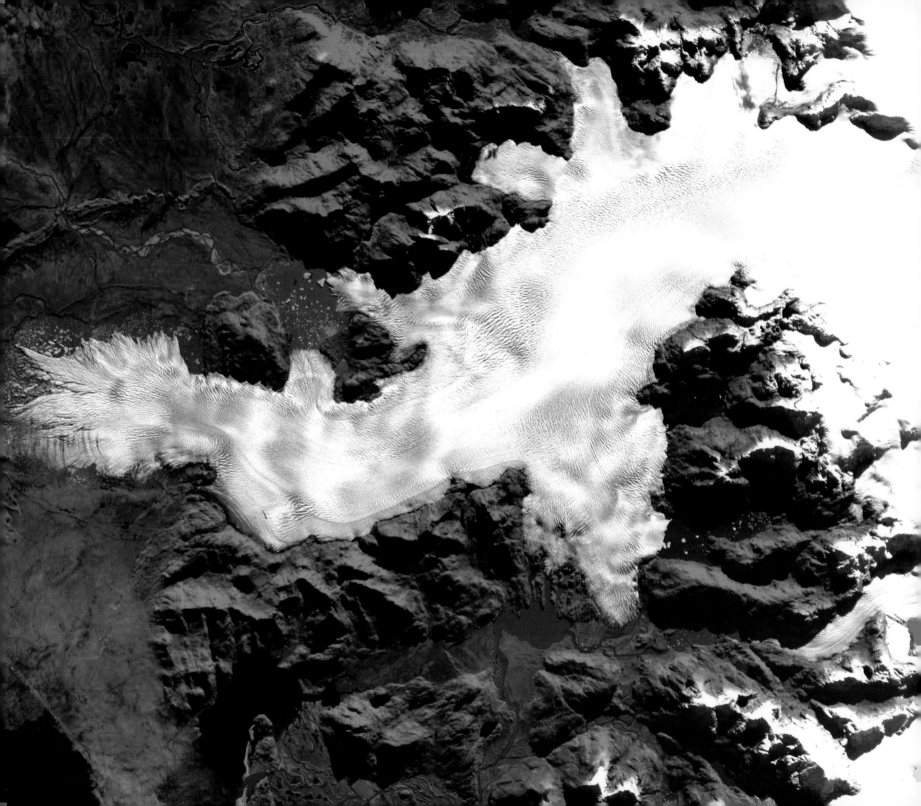

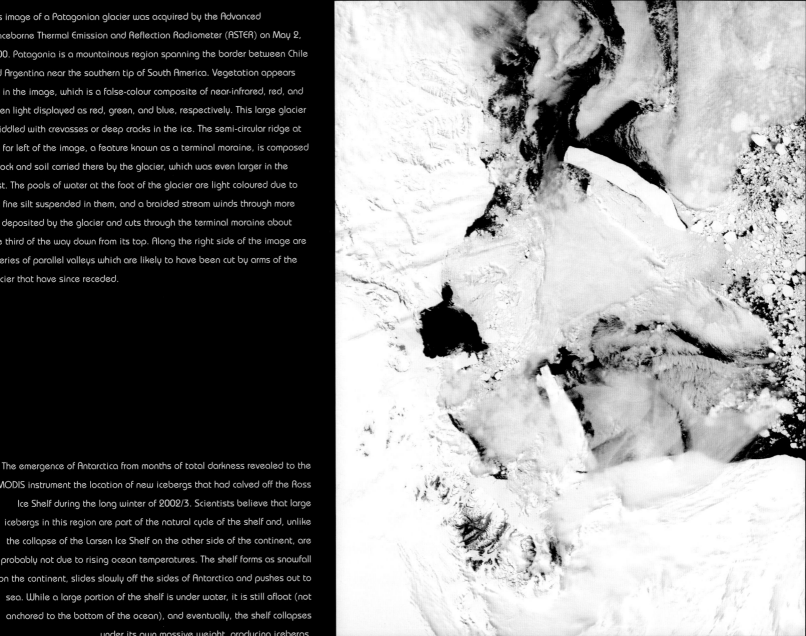

This image of a Patagonian glacier was acquired by the Advanced Spaceborne Thermal Emission and Reflection Radiometer (ASTER) on May 2, 2000. Patagonia is a mountainous region spanning the border between Chile and Argentina near the southern tip of South America. Vegetation appears red in the image, which is a false-colour composite of near-infrared, red, and green light displayed as red, green, and blue, respectively. This large glacier is riddled with crevasses or deep cracks in the ice. The semi-circular ridge at the far left of the image, a feature known as a terminal moraine, is composed of rock and soil carried there by the glacier, which was even larger in the past. The pools of water at the foot of the glacier are light coloured due to the fine silt suspended in them, and a braided stream winds through more silt deposited by the glacier and cuts through the terminal moraine about one third of the way down from its top. Along the right side of the image are a series of parallel valleys which are likely to have been cut by arms of the glacier that have since receded.

The emergence of Antarctica from months of total darkness revealed to the MODIS instrument the location of new icebergs that had calved off the Ross Ice Shelf during the long winter of 2002/3. Scientists believe that large icebergs in this region are part of the natural cycle of the shelf and, unlike the collapse of the Larsen Ice Shelf on the other side of the continent, are probably not due to rising ocean temperatures. The shelf forms as snowfall on the continent, slides slowly off the sides of Antarctica and pushes out to sea. While a large portion of the shelf is under water, it is still afloat (not anchored to the bottom of the ocean), and eventually, the shelf collapses under its own massive weight, producing icebergs.

| 11 December 2000 | 29 December 2000 | 5 January 2001 |

Part of the job of the Multi-angle Imaging SpectroRadiometer (MISR) instrument aboard NASA's Terra satellite is to document the movement of huge icebergs and spreading sea ice in Antarctica's Ross Sea. These natural phenomena are adversely affecting the region's penguin population, according to a study funded by the National Science Foundation. Two massive icebergs, initially designated B-15 and C-16, broke away from the Ross Ice Shelf in March 2000 and migrated west to a point northeast of McMurdo Sound, and the resulting barrier altered wind and current patterns. In addition, earlier in the same season, sea ice in the region of the main US Antarctic facility, McMurdo Station, expanded from its normal distance of 24 to 32

kilometers (15 to 20 nautical miles) north of the base to approximately 128 kilometers (80 nautical miles). The combination of icebergs and sea ice has made it difficult for entire colonies of penguins to return from their feeding grounds in the open sea to their breeding areas, and the result was expected to be a significant reduction in regional penguin populations, with one colony in danger of extinction. The images, taken between December 2000 and December 2001, depict the rapid motion of the C-16 iceberg in late 2000 and early 2001 and its subsequent stall, as well as the incursion of the B-15A iceberg, a large fragment of the original B-15 iceberg. The increase in sea ice is particularly pronounced in the final image.

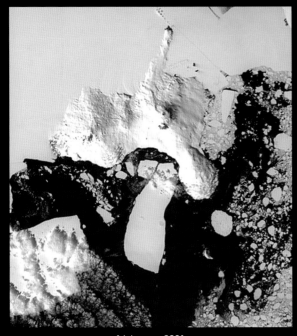

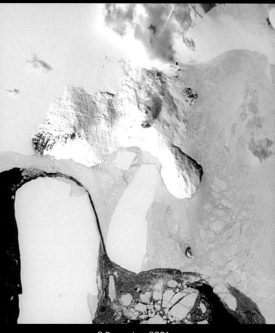

16 January 2001

15 February 2001

9 December 2001

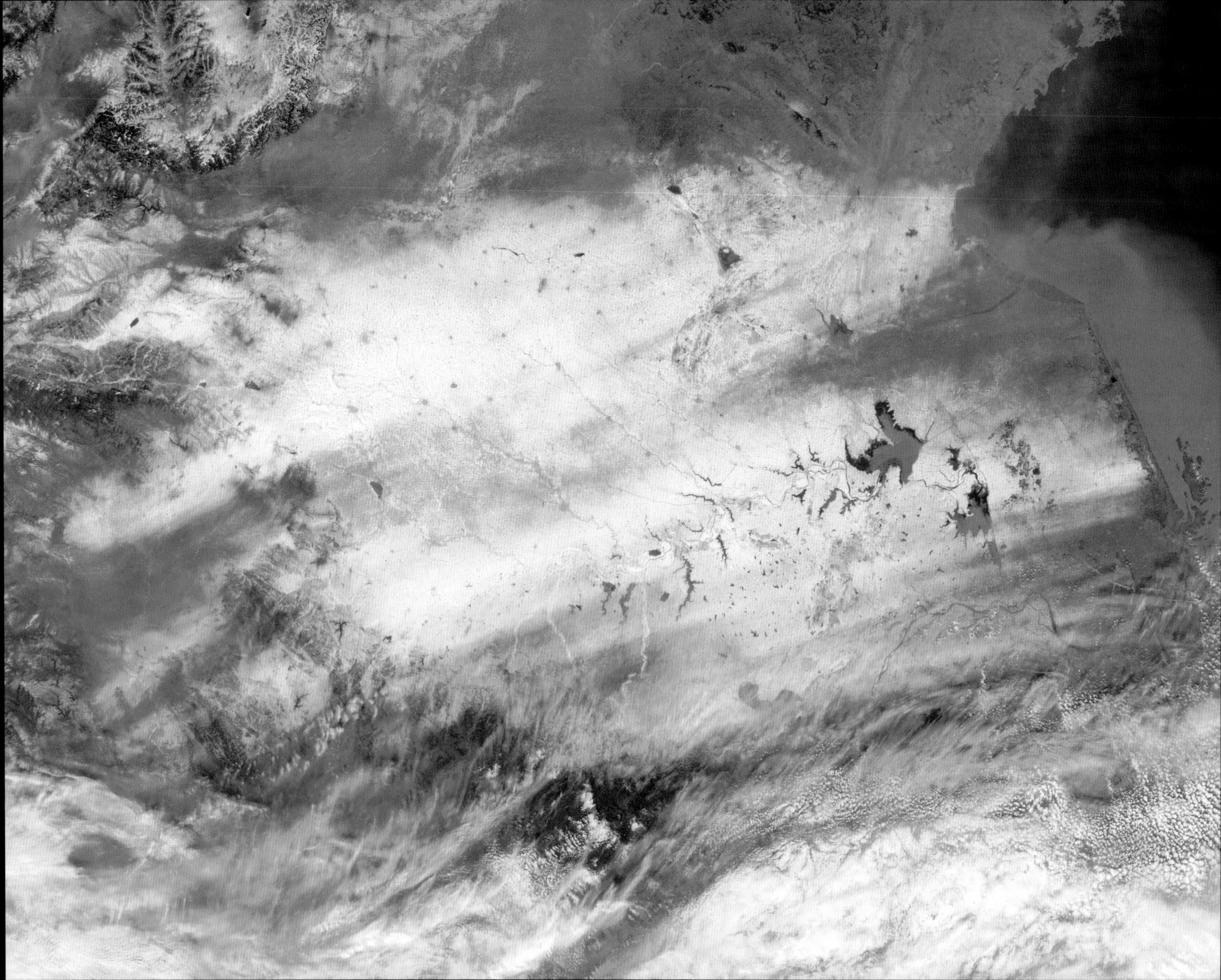

Trails of snow streak across eastern China in 2003 in this true-colour image from the Moderate Resolution Imaging Spectroradiometer (MODIS) instrument aboard NASA's Aqua satellite. In addition to the snow, which occupies the centre of the image, it is possible to see the smog that bears witness to the fact that air pollution is a persistent, year-round concern in China. Also visible in this image is the Yellow Sea (Huang Hai) in the upper right corner. Part of this appears brown and green due to the sediment carried into it by a number of rivers, including the great Yangtze.

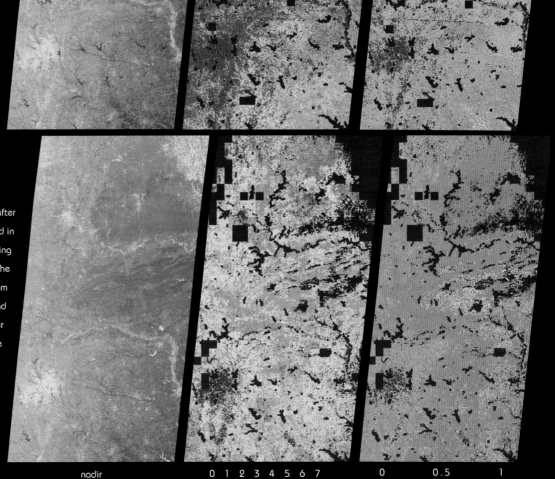

Vigorous vegetation growth in the Southern United States after heavy rains fell in April and early May, 2004, is quantified in these images and data products from the Multi-angle Imaging SpectroRadiometer (MISR) on board the Terra satellite. The images were acquired on April 1 (top set) and May 3 (bottom set), and extend through Kansas and Missouri, Oklahoma and Arkansas, and eastern Texas, with the Texas-Louisiana border at the bottom right-hand corner. The left-hand images are natural-colour views from MISR's nadir camera. The middle panels show Leaf Area Index (LAI), or the area of leaves per unit area of ground below them, as measured from above. The right-hand panels show FPAR, which is the fraction of the photosynthetically active region (PAR) of visible light (400–700 nm) absorbed by green vegetation.

nadir   0 1 2 3 4 5 6 7   0   0.5   1

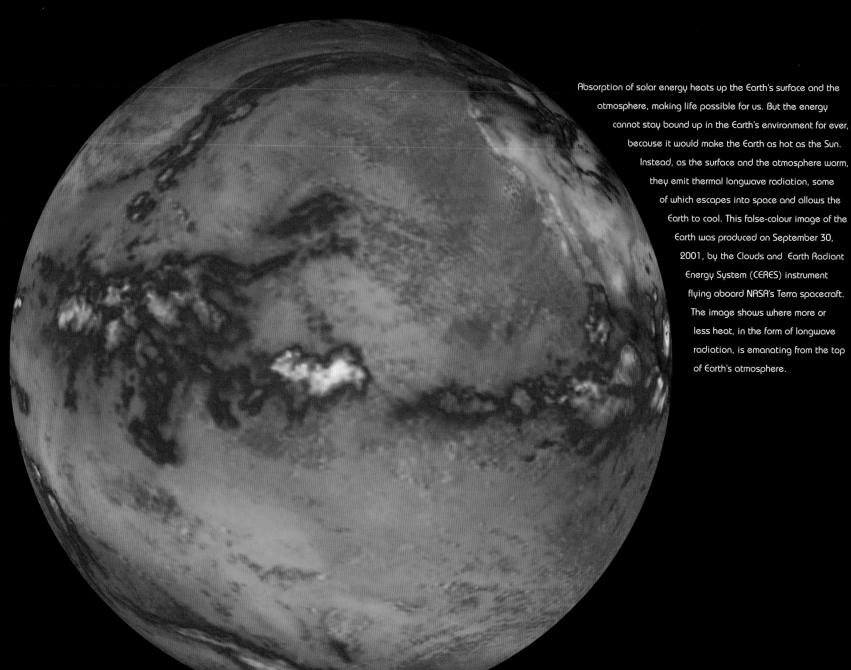

Absorption of solar energy heats up the Earth's surface and the atmosphere, making life possible for us. But the energy cannot stay bound up in the Earth's environment for ever, because it would make the Earth as hot as the Sun. Instead, as the surface and the atmosphere warm, they emit thermal longwave radiation, some of which escapes into space and allows the Earth to cool. This false-colour image of the Earth was produced on September 30, 2001, by the Clouds and Earth Radiant Energy System (CERES) instrument flying aboard NASA's Terra spacecraft. The image shows where more or less heat, in the form of longwave radiation, is emanating from the top of Earth's atmosphere.

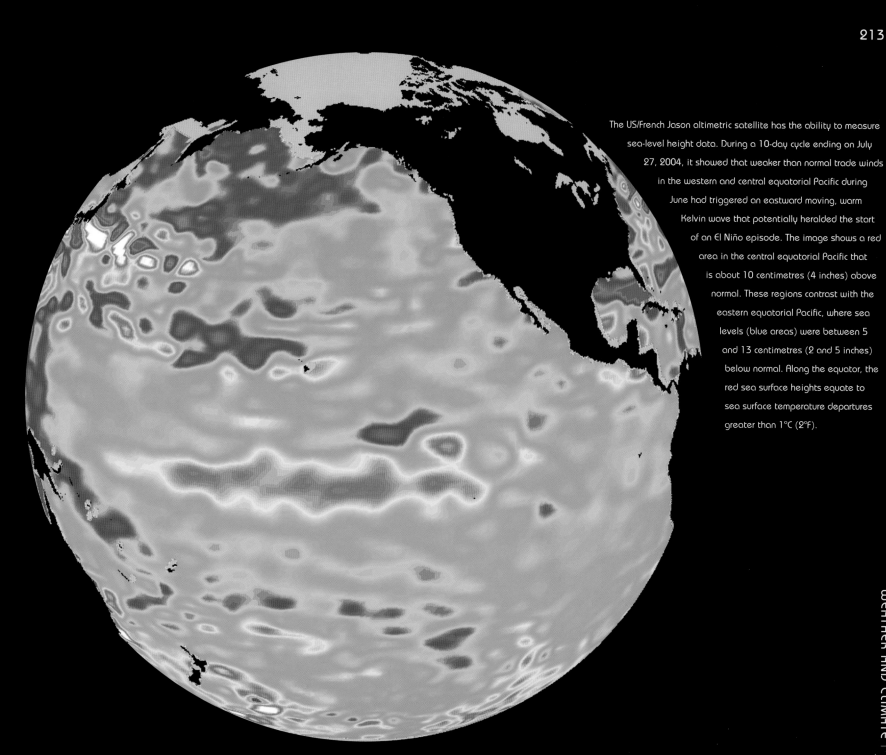

The US/French Jason altimetric satellite has the ability to measure sea-level height data. During a 10-day cycle ending on July 27, 2004, it showed that weaker than normal trade winds in the western and central equatorial Pacific during June had triggered an eastward moving, warm Kelvin wave that potentially heralded the start of an El Niño episode. The image shows a red area in the central equatorial Pacific that is about 10 centimetres (4 inches) above normal. These regions contrast with the eastern equatorial Pacific, where sea levels (blue areas) were between 5 and 13 centimetres (2 and 5 inches) below normal. Along the equator, the red sea surface heights equate to sea surface temperature departures greater than 1°C (2°F).

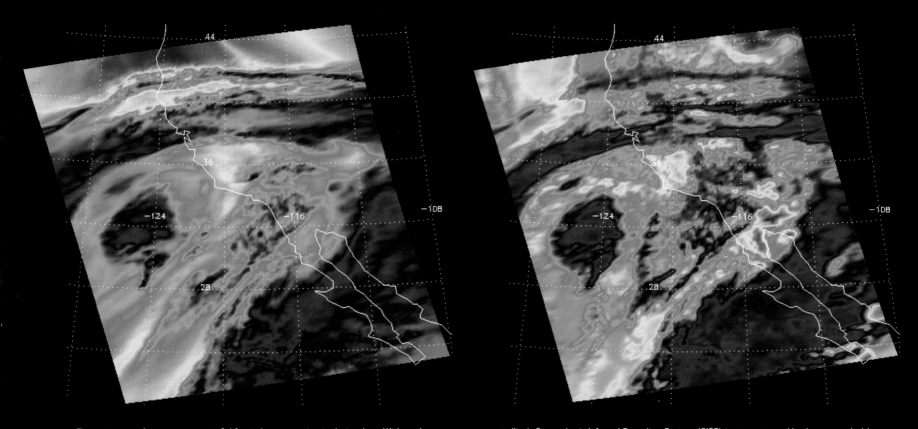

These images of a recent storm in California have meaning in their colour. With cooler areas registering as purple, and warmer areas as red, the images are a snapshot of a storm moving up from the lower latitudes. The images show a prominent squall line pointing nearly north-south that is approaching the coast, and a large isolated cloud formation almost due west. Both features have high cold cloud tops, according to the bluer images, left, achieved by the Aqua satellite's Atmospheric Infrared Sounding System (AIRS) instrument, and both were probably a major source of intense rainfall. The AMSU-A microwave sensor on the satellite, meanwhile revealed the warm land surface and the moisture below the cloud tops, right. With its visible, infrared, and microwave detectors, the AIRs experiment on Aqua can provide a three-dimensional look at Earth's weather, even with heavy clouds.

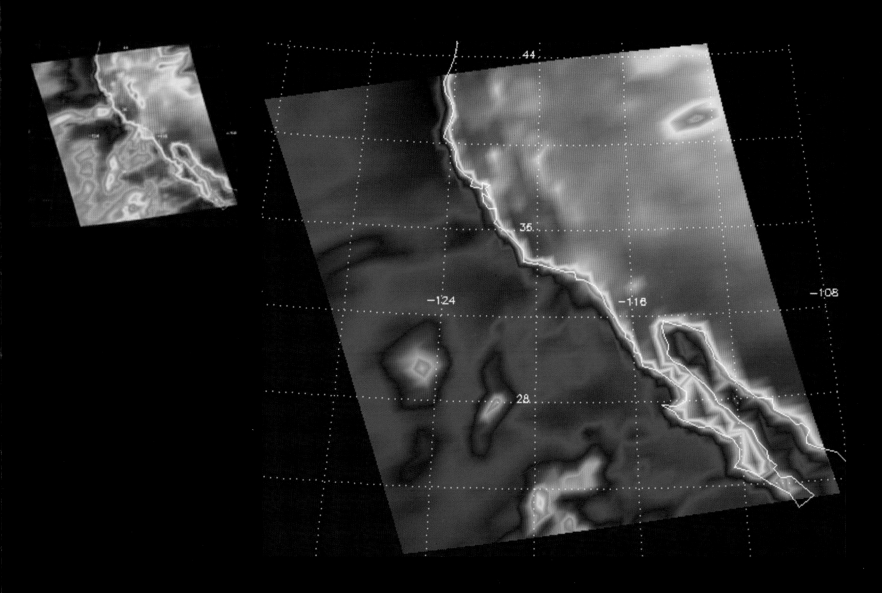

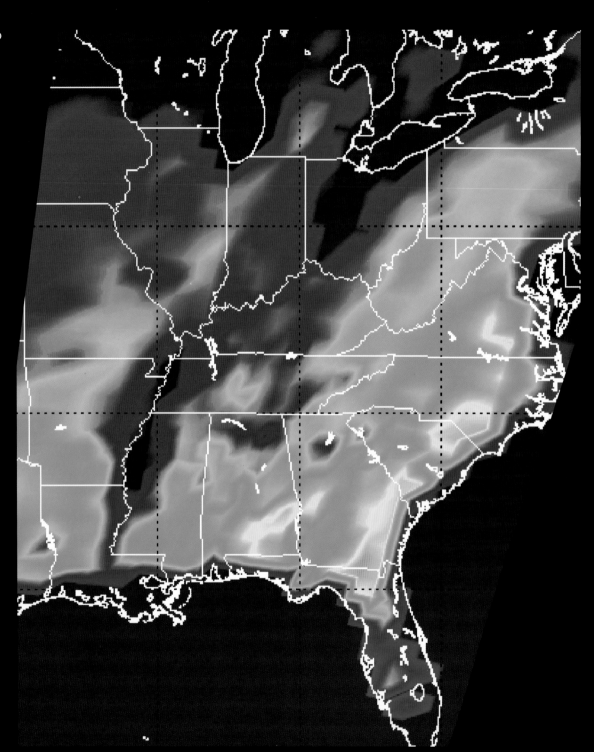

While pictures from other instruments on board the Aqua satellite showed no significant weather systems over the southeastern United States on September 12 and September 28, 2002, the microwave component of the Atmospheric Infrared Sounder Experiment (AIRs) showed a striking difference. This is primarily due to flooding that had occurred after Tropical Storm Isidore. Water has a very low surface emissivity at this frequency, and that causes surface water to appear very cold (even though it is not). Land appears relatively warm (well above freezing – 273 Kelvin, even at night as seen in these images), but if there is standing water, the apparent temperature drops accordingly. The image to the right, taken just about a day after the remnants of Isidore passed over the southeast, shows heavy flooding along the Mississippi, especially in the states of Mississippi and Tennessee, but other states are also affected. The spatial resolution of the AMSU-A instrument is relatively large (each measurement spot is about 25 miles, or 40 kilometres, in diameter), but the enormous thermal contrast in the microwave between land and water makes even small flooded areas stand out.

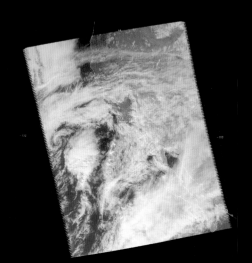

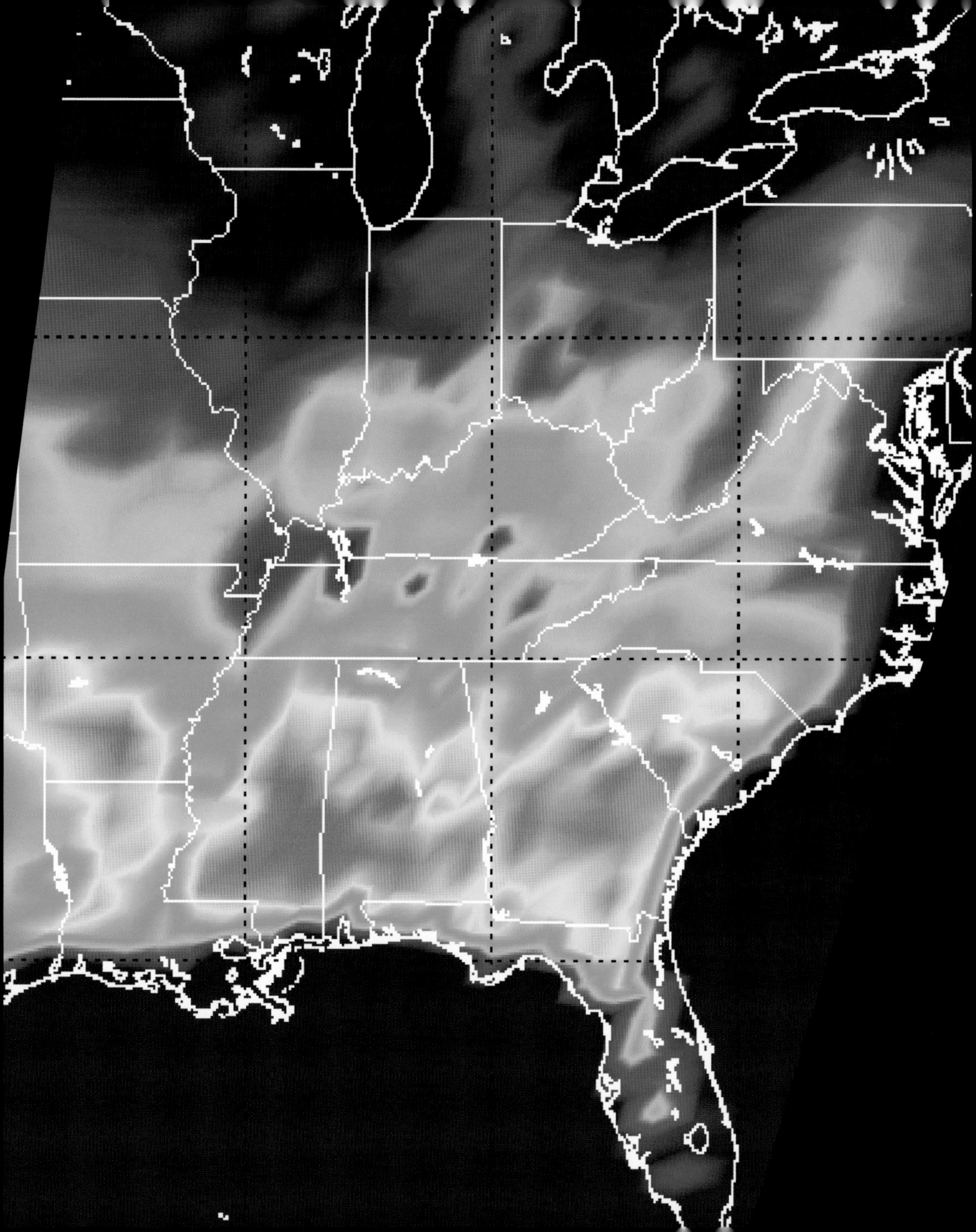

This digital mosaic of clouds around the Earth was taken
by the Environmental Science Services Administration
(ESSA) 5 satellite on September 14, 1967. It
shows more than a dozen storm areas, including
hurricanes Beulah, Dora, Chloe, Monica, and
Nannette. The picture signals, received in
analogue form from the satellite, were
sampled at short time intervals, and
numbers were assigned according to
the brightness indicated by the basic
signal. The computer then located
each digital value and precise
map location. The computer tape
containing this information was then
displayed on a television-like tube
(kinescope) and photographed, and
the result was this mosaic.

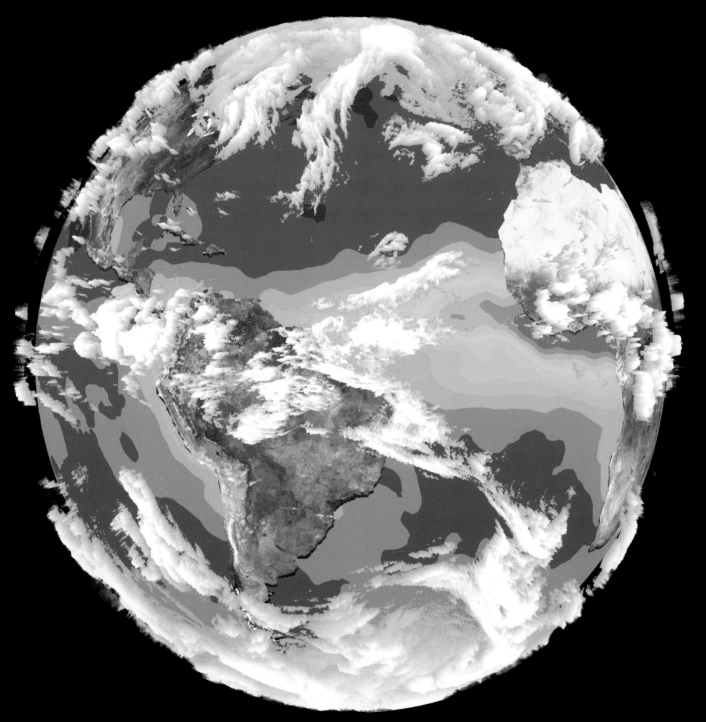

Satellite data and images such as those presented in this image of Earth give scientists a more comprehensive view of the Earth's interrelated systems and climate. Four different satellites contributed to the making of this image. Sea-viewing Wide Field-of-view Sensor (SeaWiFS) provided the land image layer and is a true colour composite of land vegetation for cloud-free conditions from September 18 to October 3, 1997. Each red dot over South America and Africa represents a fire detected by the Advanced Very High Resolution Radiometer. The oceanic aerosol layer is based on National Oceanic and Atmospheric Administration (NOAA) data and is caused by biomass burning and windblown dust over Africa. The cloud layer is a composite of infrared images from four geostationary weather satellites: NOAA's GOES 8 and 9, the European Space Agency's METEOSAT, and Japan's GMS 5.

WEATHER AND CLIMATE |

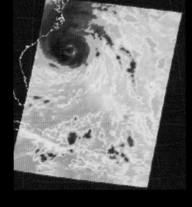

These two false-colour images show Hurricane Isabel viewed by the AIRS and AMSU-A instruments on board the Aqua satellite on the morning of September 18, 2003. Isabel was due to come ashore within 12 hours, bringing widespread flooding and destructive winds. In the image above, data retrieved by the AIRS infrared sensor shows the hurricane's eye as the small ring of pale blue near the upper left corner of the image. The dark blue band around the eye shows the cold tops of hundreds of powerful thunderstorms. The second image (right) shows the geographical distribution and total amount of atmospheric water vapour associated with Isabel as inferred by AIRS and AMSU-A. Very humid areas appear deep red and surround the storm's eye in the ring of thunderstorms.

Cloud-top radiance and height characteristics of Hurricane Isabel are shown in these data products and animations from the Multi-angle Imaging SpectroRadiometer (MISR) aboard the Terra satelitte. MISR observes the Earth continuously from pole to pole, and every 9 days views the entire globe between 82 degrees north and 82 degrees south latitude. Isabel was upgraded to hurricane status a few hours after the top images in this set were acquired on September 7, 2003. By the time the bottom panels were acquired on September 11, Isabel was a strengthening category 4 hurricane, centred about 900 kilometers east-northeast of the Leeward Islands. To the left are radiance images from MISR's vertical-viewing (nadir) camera, in the centre are cloud-top height fields, and the right-hand panels provide retrieved local albedo values. Albedo is a function of the amount of sunlight reflected back to space divided by the amount of incident sunlight.

nadir

0   5   10   15   20

height in kilometers

0   0.5   1

a bedo

WEATHER AND CLIMATE

After causing widespread destruction on Puerto Rico, Haiti and the Dominican Republic, Hurricane Jeanne was weakened to Tropical Storm status for several days before it regained strength over the Bahamas as a Category 2 hurricane. When Jeanne made landfall over US territory on September 26, it was the fourth major hurricane of the 2004 Atlantic hurricane season to strike Florida. These visualizations of Hurricane Jeanne on September 24 were captured by the Terra satellite's Multi-angle Imaging SpectroRadiometer (MISR). The still panels include a natural colour view from MISR's 26-degree forward-viewing camera (left) and a two-dimensional map of cloud-top heights (right).

26° forward

0   10   20

height in kilometers

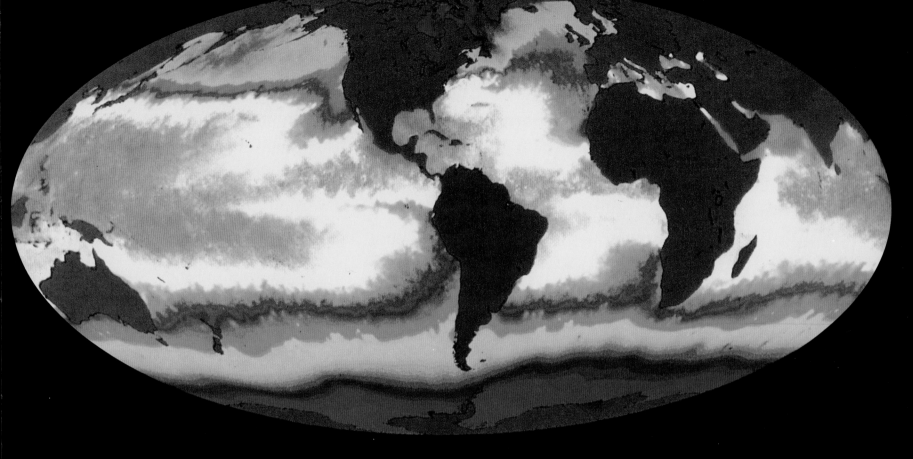

This illustration of Earth's sea surface temperature was obtained from two weeks of infrared observations by the Advanced Very High Resolution Radiometer (AVHRR), an instrument on board the NOAA-7 satellite during July 1984. Temperatures are colour coded, with red being warmest and decreasing through oranges, yellows, greens, and blues. Temperature patterns seen in this image are the result of many influences, including the circulation of the ocean, surface winds, and solar heating. The image indicates a large pool of warm water in the Western Pacific and a tongue of relatively cold water extending along the Equator westward from South America.

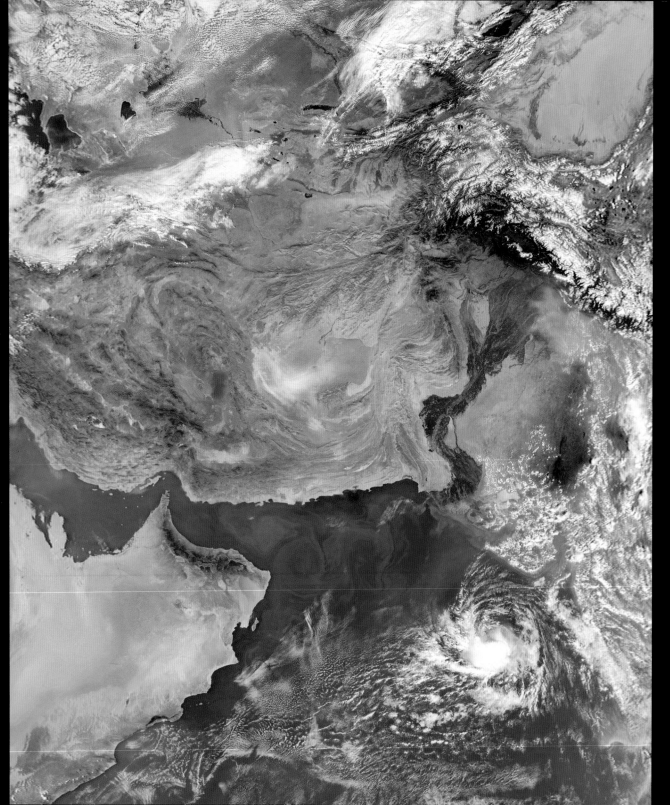

On October 6, 2004, the Sea-viewing Wide Field-of-view Sensor aboard the OrbView-2 satellite captured these two images showing high concentrations of phytoplankton in the Arabian Sea. The chlorophyll that the plants use to convert light to food tints the water green in the natural colour image (left). The phytoplankton are growing in large swirls that follow the eddies and currents of the surface water. In the image opposite, ocean chlorophyll concentrations are shown. Not surprisingly, concentrations appear to be highest near the coast where upwelling makes nutrients more available.

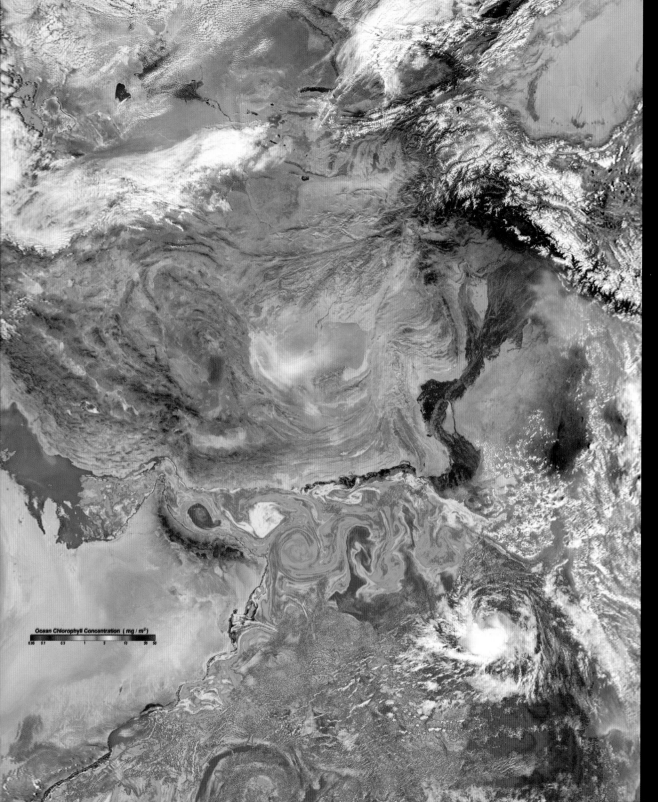

Ocean Chlorophyll Concentration (mg / m³)

0.05   0.1   0.3   1   3   10   30   50

On November 18, 2003, the Advanced Spaceborne Thermal Emission and Reflection Radiometer (ASTER) on the Terra satellite acquired this image of the Old Fire/Grand Prix fire east of Los Angeles. The image was processed by NASA's Wildfire Response Team and was then sent to the United States Department of Agriculture's Forest Service Remote Sensing Applications Center (RSAC), whose role it is to provide interpretation services to Burned Area Emergency Response (BAER) teams to assist in mapping the severity of the burned areas. The image combines data from the visible and infrared wavelength regions to highlight the burned areas.

With sustained winds of 257 kph (160mph) and gusts of up to 298 kph (185 mph), Super Typhoon Ma-On was situated due south of Japan on October 8, 2004. The eye of the storm was located about 1,000 kilometres (621 miles) southwest of Tokyo and was moving north-northeastwards at about 40 kph (25 mph). The true-colour image was acquired on by the MODIS instrument aboard NASA's Aqua satellite.

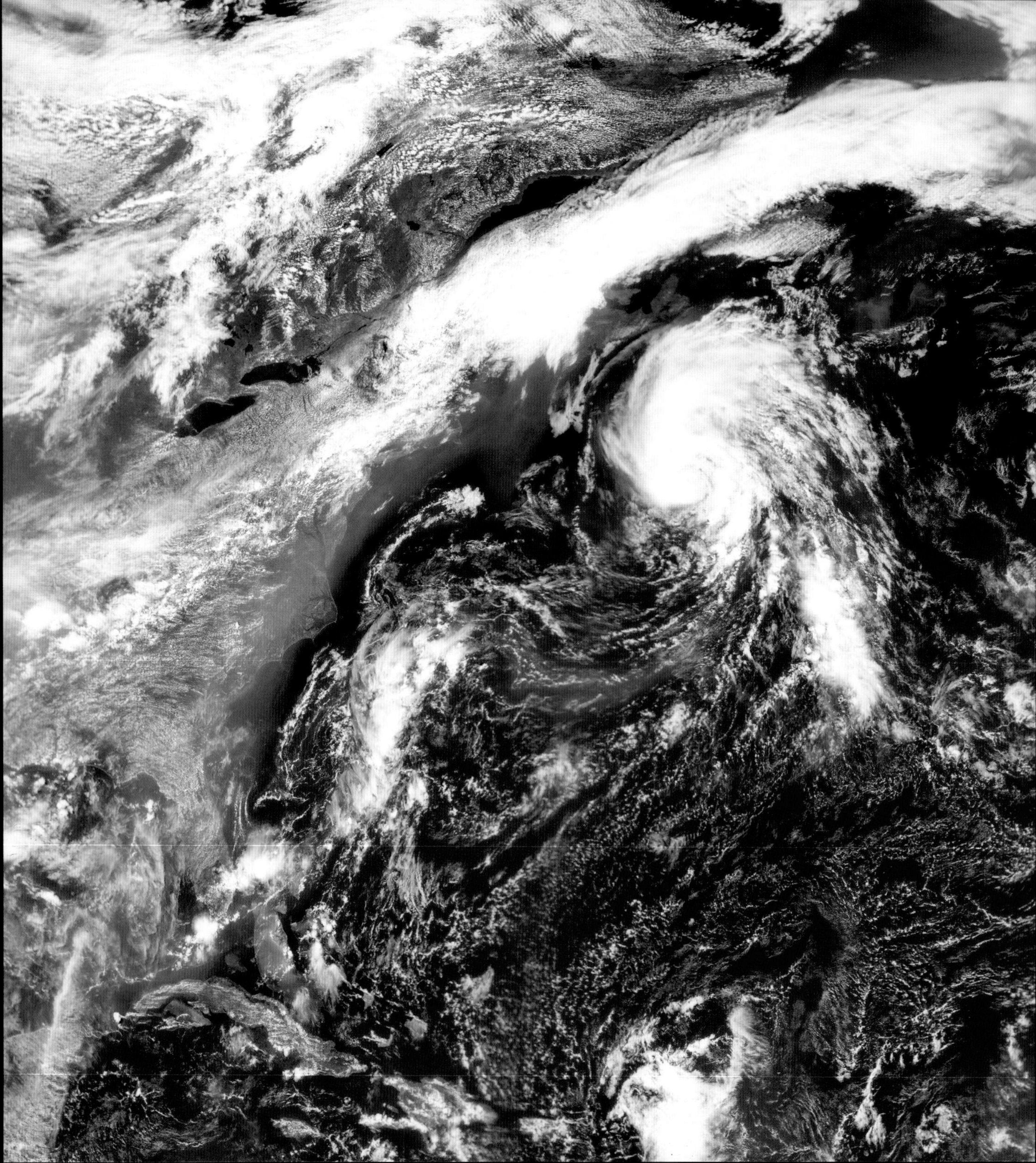

Tropical Storm Dean was spinning
to the southeast of Nova Scotia
in this SeaWifS satellite image
captured on August 27, 2001.
The image also shows bright
aquamarine water over the Grand
Banks of Newfoundland to the
northeast of the storm. To the
west of the storm, a thick band of
aerosol hangs over the east coast
of the United States, while dust
from Africa is visible in the lower
right corner of the image.

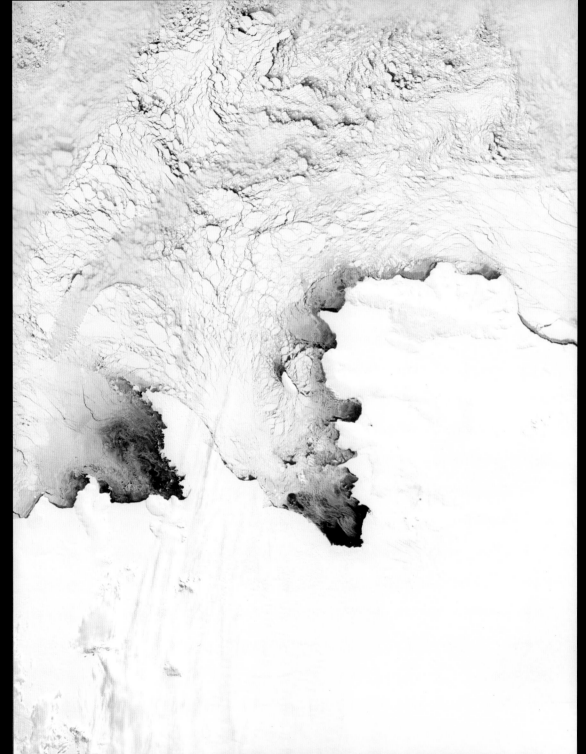

This true-color MODIS image shows the Walgreen
Coast, Eights Coast, and Amundsen Sea. It was
captured on October 23, 2001.

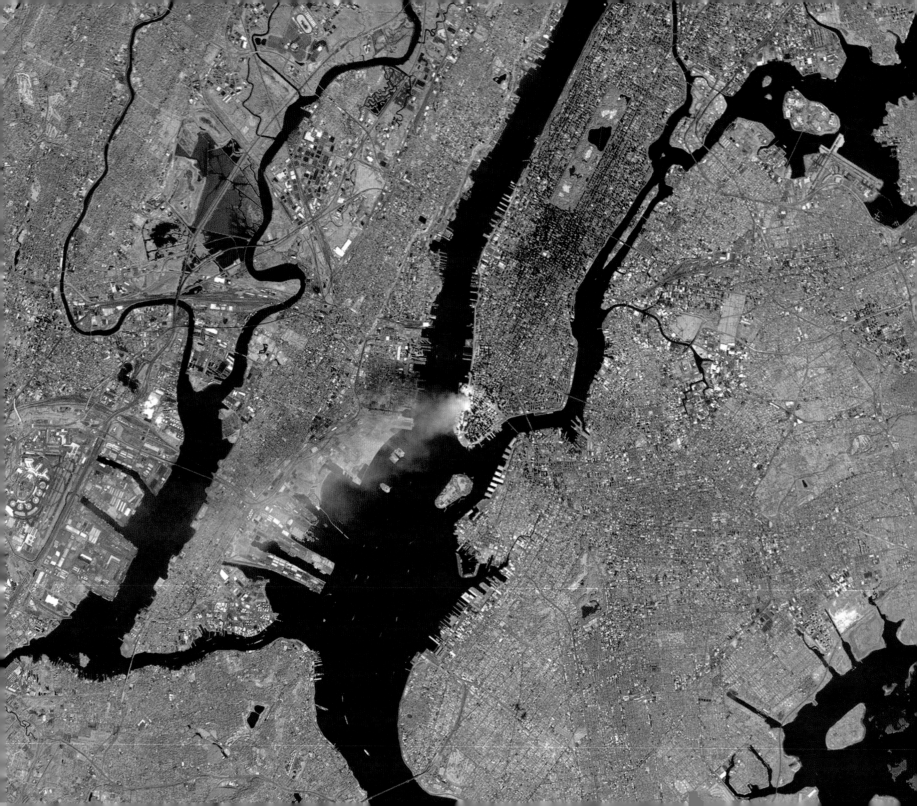

DRAMATIC INCIDENTS

On June 26, 2003 the Advanced Spaceborne Thermal Emission and Reflection Radiometer (ASTER) on board the Terra satellite acquired this image of the Aspen fire burning out of control north of Tucson, Arizona. As of that date, the fire had consumed more than 109 square kilometres (27,000 acres) and destroyed more than 300 homes, mostly in the resort community of Summerhaven. Data such as this is used by NASA's Wildfire Response Team and the US Forest Service to assess the intensity of the burn, and to help with firefighting efforts.

PREVIOUS PAGE
This true-colour image of the aftermath of the attack on the World Trade Centre in New York was taken by the Enhanced Thematic Mapper Plus (ETM+) aboard the Landsat 7 satellite on September 12, 2001, at roughly 11:30 am. Eastern Time.

Mount Etna appears to be producing a smoke plume in this 2000 SeaWiFS image. Also visible in this scene are some interesting wave patterns on the eastern side of Sardinia. These are probably caused by differing amounts of sunglint (sun reflecting off a water surface) from a sea surface with variable roughness and wave direction.

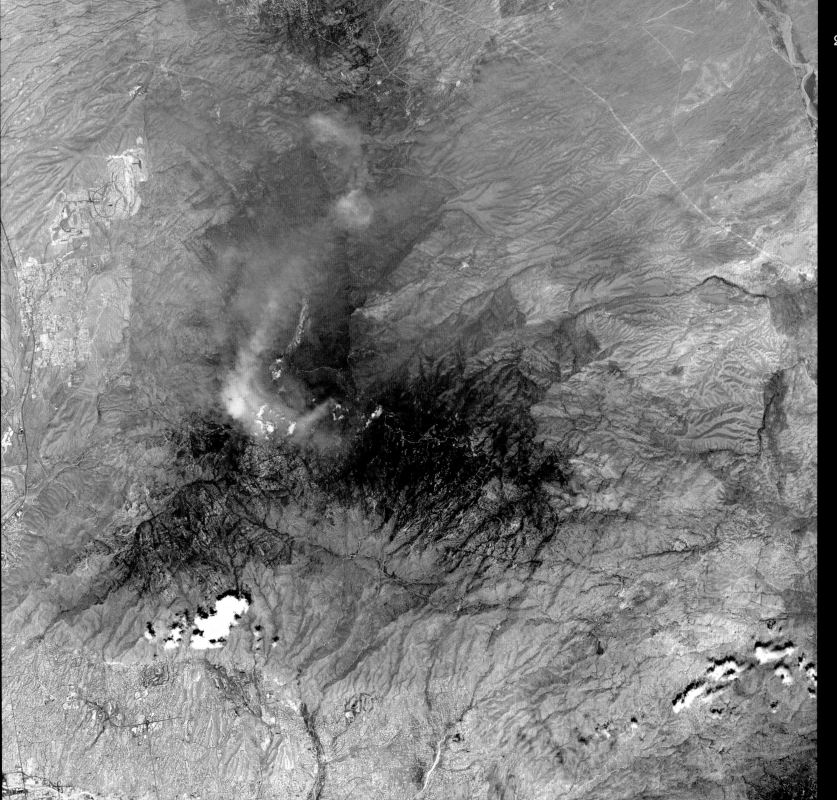

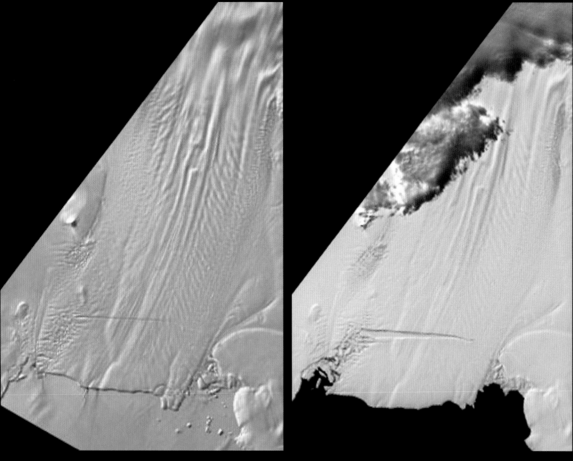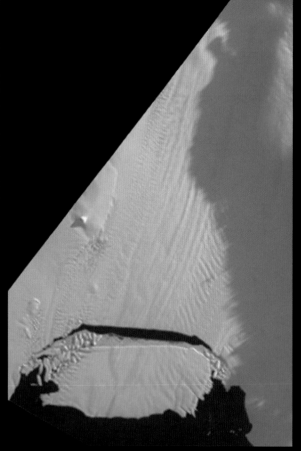

16 September 2000   4 November 2001   12 November 2001

The image set shows three views of Pine Island Glacier acquired by the
Terra satellite using its Multi-angle Imaging SpectroRadiometer's (MISR)
vertical-viewing (nadir) camera. The first was captured in late 2000,
early in the development of the crack. The second and third views were
acquired in November 2001, just before and just after the new iceberg
broke off. The existence of the crack took the glaciological community
by surprise, as did the rapid rate at which the crack spread, since the
rift was not predicted to reach the other side of the glacier before
2002. However, the iceberg detached much sooner than anticipated,
and the last 10-kilometre (6-mile) segment that was still attached to
the ice shelf snapped off in a matter of days.

As Antarctica emerges from months of darkness, new icebergs are revealed by a MODIS instrument aboard a passing satellite in 2003. The fact that some are on the move was revealed by MODIS tracking the area during several passes. Along the western portion of the image lie the Trans-Antarctic Mountains, while in the centre is the Ross Sea and banked up against the Antarctic coastline is the vast Ross Ice Shelf (bottom right).

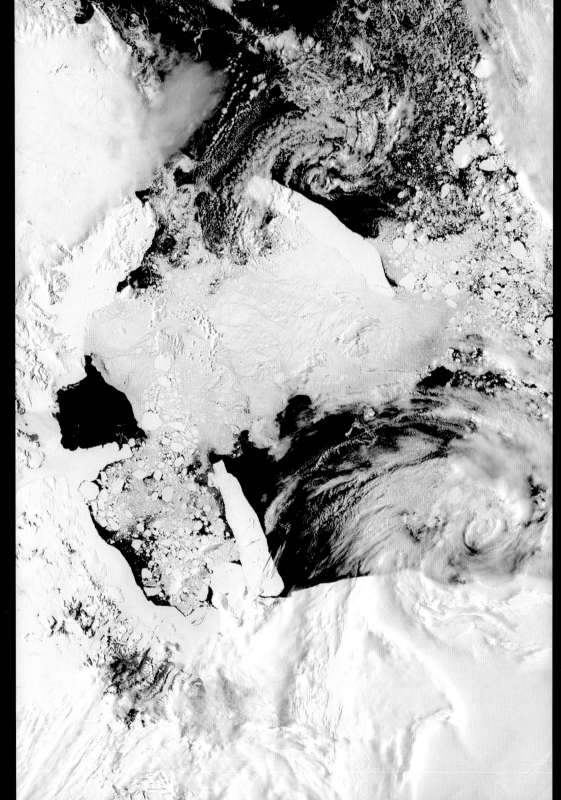

These two images that show the breakup of the northern section of the Larsen B ice shelf in March 2002 were produced by the MISR instrument aboard the Terra satellite. The collapse is thought to have been accelerated by warm summer temperatures which caused melt water to fill crevasses along the landward side of the Larsen shelf, leading to intensified pressures within the sheet structure. In the left-hand view, spectral variations across the scene are highlighted by using near-infrared, red and blue data from MISR's nadir (vertical-viewing) camera, and here the ice within the disintegrating ice shelf appears vibrant blue. The right-hand view was created from data obtained from three different view angles and only one colour channel, and this was combined to create a multi-angle composite. This image displays red-band data from MISR's 46-degree forward, nadir, and 46-degree backward-viewing cameras as red, green and blue, respectively.

SeaWifS captured this image in 2001 showing what appears to be large quantities of dust blowing southward out of Iceland, the island itself being mostly hidden by clouds. Sea ice is visible along the Greenland

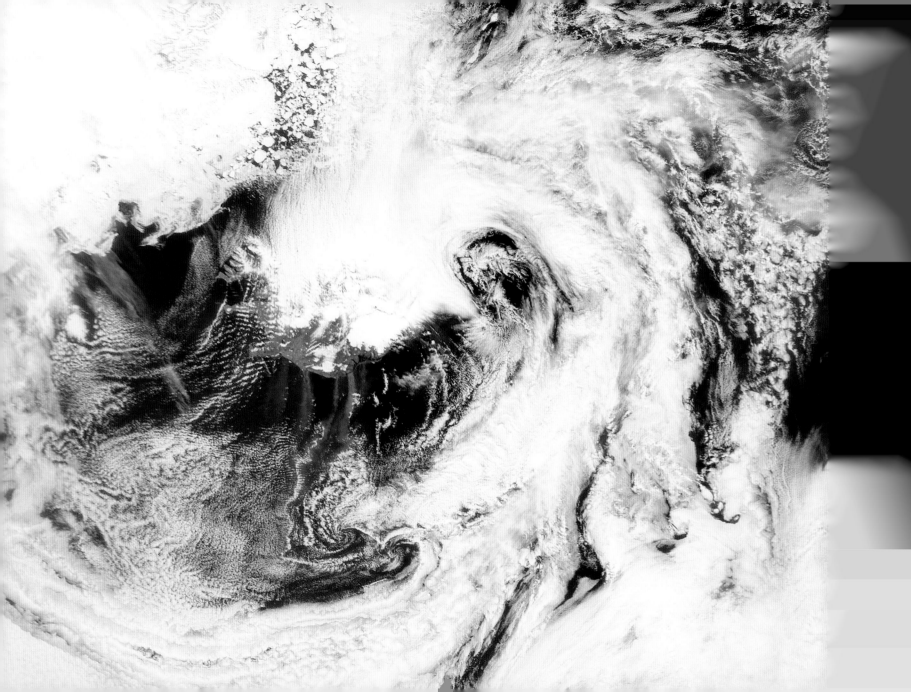

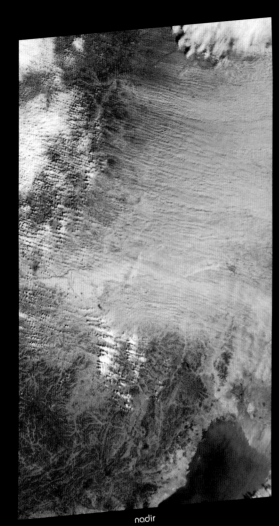

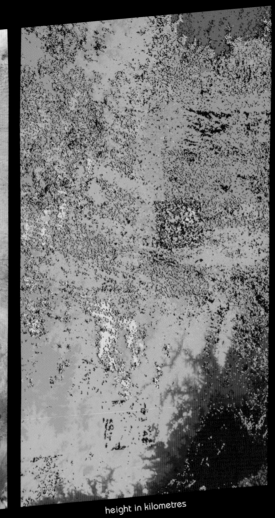

nadir

height in kilometres

height in kilometres

The MISR instrument on board the Terra satellite captured these views of the dust and sand that swept over northeast China on March 10, 2004. Information on the height of the dust and an indication of the probable dust source region are provided by these images, which include a natural-colour snapshot from MISR's nadir camera (left), a stereoscopically-retrieved height field (centre) and a map of terrain elevation (right). A decrease in spring vegetation coverage in central and eastern Inner Mongolia has been suggested to be a major contributor to spring dust storms over northern China.

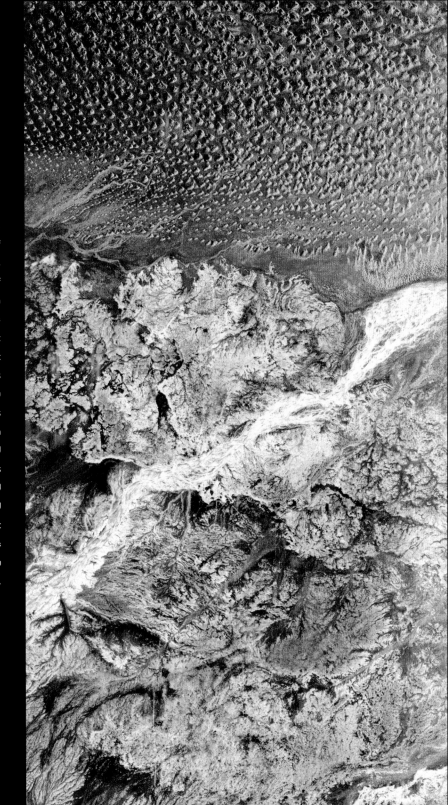

This is a radar image of the region around the site of the lost city of Ubar in southern Oman, on the Arabian Peninsula, a site that was discovered in 1992 with the aid of remote sensing data. Archaeologists believe Ubar existed from about 2800 BC to about 300 AD and was a remote desert outpost where caravans were assembled for the transport of frankincense across the desert. This image was acquired on orbit 65 of the space shuttle Endeavour on April 13, 1994 by the Spaceborne Imaging Radar C/X-Band Synthetic Aperture Radar (SIR-C/X-SAR), and it covers an area about 50 kilometres by 100 kilometres (31 miles by 62 miles). The radars from this instrument illuminate Earth with microwaves, allowing detailed observations at any time, regardless of weather or sunlight conditions, and the multi-frequency data will be used by the international scientific community to better understand the global environment and how it is changing.

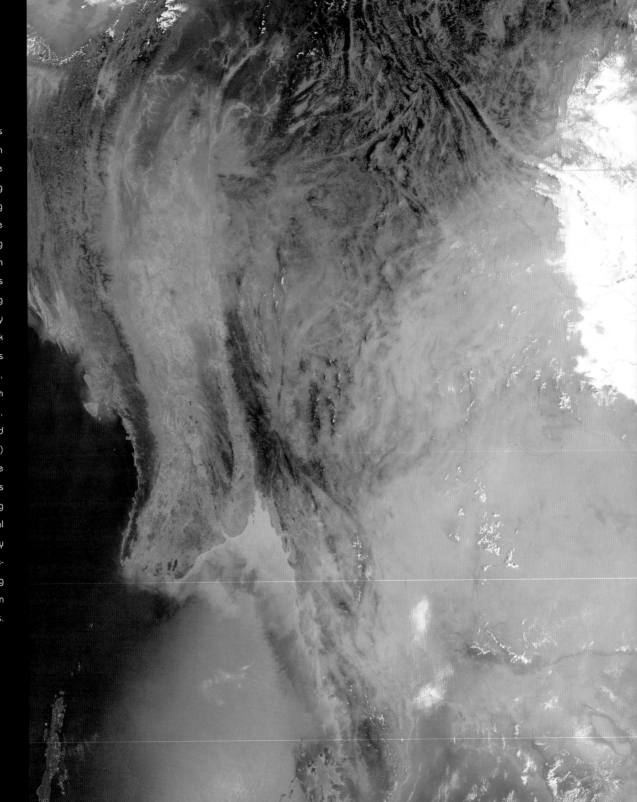

Across Southeast Asia, the biomass burning season is in high gear in mid-March 2004. This is the time of year for agricultural burning – clearing farmland and renewing farmland and rangeland to prepare for the coming spring growing season. In this MODIS image from March 15, 2004, scores of fires were detected by the sensor during a Terra satellite overpass, and they are marked with red dots. A thick blanket of blue-grey smoke hangs over the right half of the image, shrouding Cambodia (bottom right), Thailand, (to the northwest), Laos (northeast of Thailand), and parts of southern China (top right) and Myanmar (to the west). The widespread nature of the fires suggests that they are being set intentionally for agricultural purposes. Though not necessarily immediately hazardous, such large-scale burning can have a strong impact on weather, climate, human health, and natural resources.

During the 2003 fire season, blazes in the taiga forests of Eastern Siberia were part of a vast network of fires across Siberia and the Russian Far East, northeast China and northern Mongolia. Fires in Eastern Siberia have been increasing in recent years, and the 2003 spring and summer seasons are the most extensive recorded in over 100 years. Overall, the Russian Federation experienced a record-setting fire year, with over 55 million acres burnt by early August. This set of three images from the MISR instrument on board the Terra satellite illustrate the extent and height of smoke from numerous fires in the Lake Baikal region on June 11, 2003. The left and centre panels are natural-colour views from MISR's vertical-viewing (nadir) and 70-degree forward-viewing cameras, respectively, while on the right is a map of stereoscopically retrieved heights for features exhibiting sufficient spatial contrast.

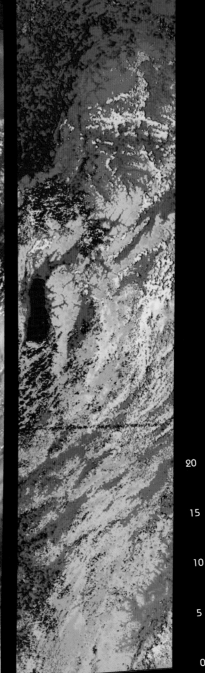

20

15

height in kilometers

10

5

0

DRAMATIC INCIDENTS

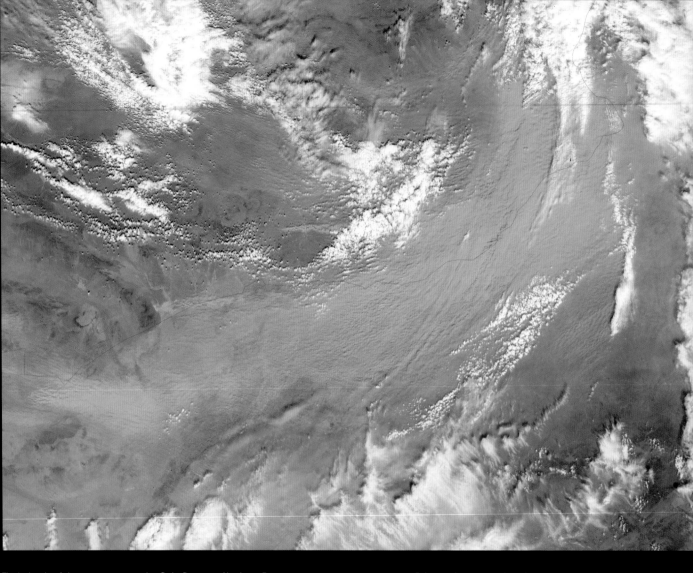

Thick clouds of dust sweep across the Gobi Desert in Northern China and Southern Mongolia in this MODIS image taken by the Terra satellite on March 27, 2004. According to China's official news agency, about 70 million people in 11 provinces were affected by sandstorms on March 27 and 28. The worst of the storm was concentrated here, in Inner Mongolia, where yellow sand blew for sixteen hours. Mongolia, top, was also affected by the storm.

Saharan dust is seen here being swept north over Libya and across the Mediterranean Sea to Sicily and Greece. Such storms are common, as hot air over the vast African desert is pulled towards the cooler winter air in the north. The strong winds that result carry Saharan dust into the Mediterranean and across Europe. The MODIS instrument on the Aqua satellite captured this storm over Libya, right, and Tunisia, left, on March 27, 2004.

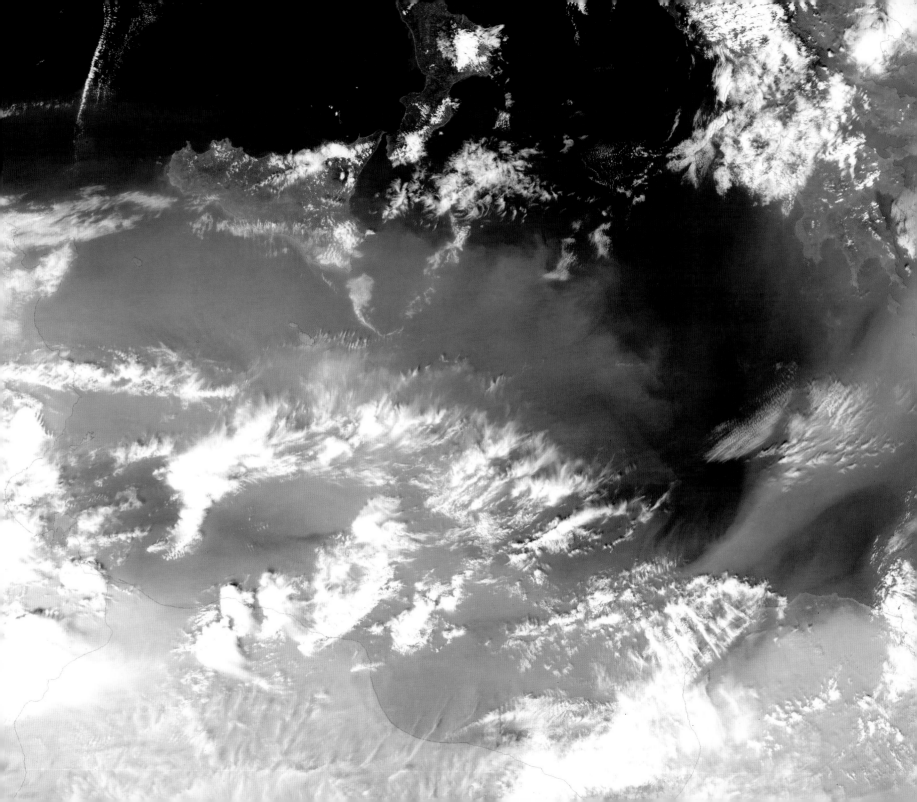

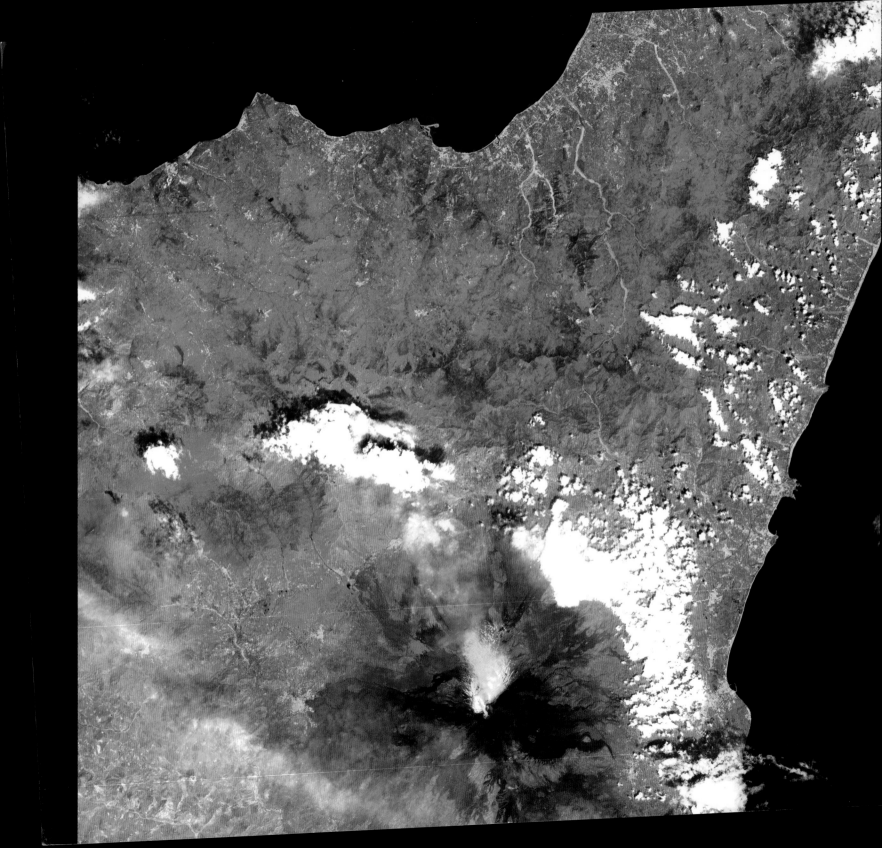

Located near Sicily, Mount Etna is Europe's most active volcano and one of the world's largest at about 3,350 metres (10,991 feet) tall. It is covered by historic lava flows as old as 300,000 years. This false colour view was acquired by the ASTER instrument aboard the Terra satellite, using a combination of ASTER's visible and near-infrared channels. The blue-white areas on the north slope are snow; the dark brown and black areas radiating out from Etna's caldera are exposed rock surfaces from previous lava flows. The deep red hues around the edges of the image indicate the presence of vegetation.

When this photograph was taken by the Expedition 2 crew aboard the International Space Station on July 22, 2001, the Italian city of Catania (in shadow, around 25 kilometres SSE of the volcano) was covered by a layer of ash, and an ash cloud was reported to have reached a height of 5.2 kilometres. Plumes from two sources are visible here – a dense, darker mass from one of the three summit craters of Mount Etna and a lighter, lower one. The record of historical volcanism of Mount Etna is one of the world's longest, dating back to 1,500 BC. Two styles of activity are typical: explosive eruptions, sometimes with minor lava flows, from the summit craters, and flank eruptions from fissures.

Thousands of fires burning in Southeast Asia were blanketing the region with blue-grey smoke when this MODIS image was captured by the Aqua satellite on March 27, 2004. Although agricultural fires are not necessarily immediately hazardous, such large-scale burning can have a strong impact on weather, climate, human health, and natural resources.

Mayon volcano is the most active volcano in the Philippines, located 325 kilometres (202 miles) southeast of Manila. A near-perfect cone, the steep, forested slopes look like a bull's eye when viewed from above. Its circular footprint is about 16 kilometres (10 miles) in diameter. This photograph was taken from the Space Shuttle on April 8, 1997 as Mayon spouted steam from the summit. The lighter (non-forested) regions that radiate from the summit to the southern slopes are flows from eruptions over the past 25 years.

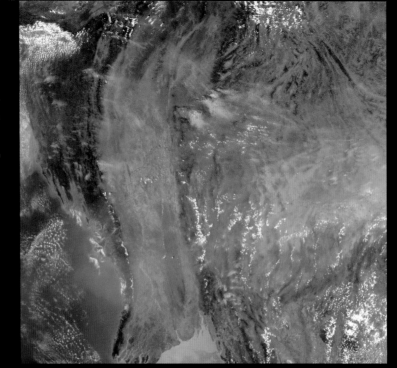

Using data from NASA's Aqua satellite, agency scientists have discovered that heavy smoke from burning vegetation inhibits cloud formation. The research suggests the cooling of global climate by pollutant particles or 'aerosols' may be smaller than previously estimated. During the August–October 2002 burning season in South America's Amazon River basin, scientists observed that cloud cover decreased from about 40 percent in clean-air conditions to zero in smoky air. This upset the theory that aerosols, such as smoke particles, served to cool the planet by shading the surface, either directly, by reflecting sunlight back toward space, or indirectly, by making clouds more reflective.

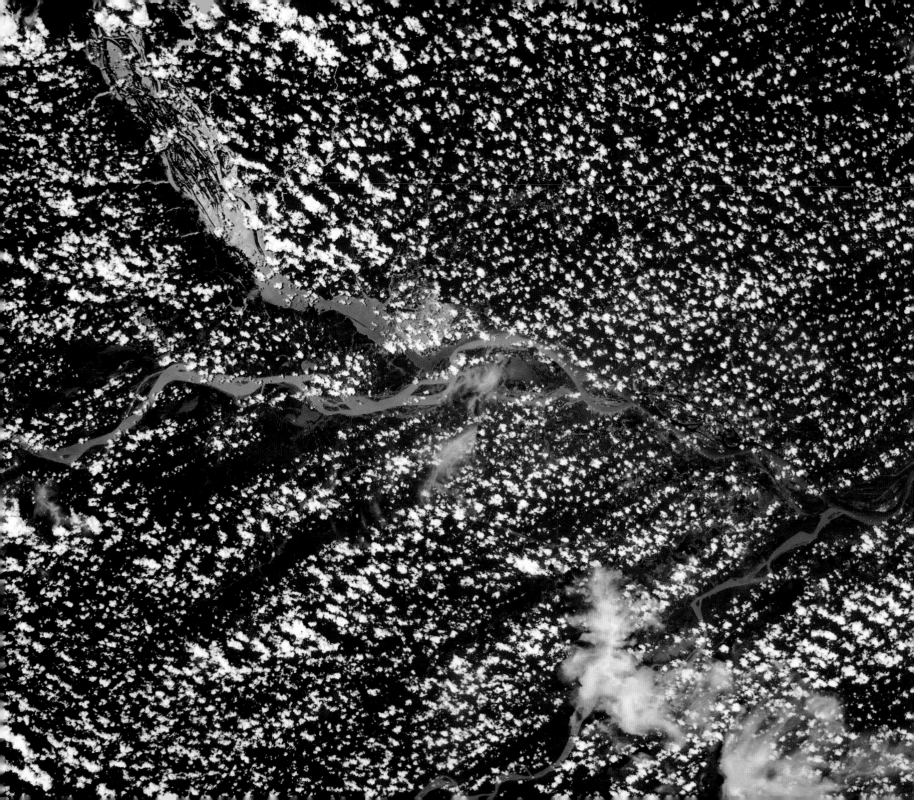

On December 26, 2004, tsunamis swept across the Indian Ocean, spawned by a massive 9.0 earthquake off the coast of Sumatra. The island nation of Sri Lanka suffered some of the worst devastation. DigitalGlobe's Quickbird satellite captured an image of Kalutara Beach on January 1, 2004, nearly a year before the disaster, top left. The other three views were taken on the day the tsunami struck. The most dramatic picture, below right, was taken at 10:20am local time — about an hour after the first in the series of waves had hit.

The Indonesian town of Banda Aceh suffered terrible damage in the tsunami of December 26, 2004. The picture, above, from DigitalGlobe's Quickbird satellite on June 23, 2004 shows the scene before disaster struck. The contrast with the Quickbird image taken on December 28, 2004 couldn't be greater. Vast areas of the town have simply been washed away, with vast loss of life.

Gleebruk Village in Indonesia is seen before the tsunami
on April 12, 2004, left, and again on January 2, 2005.
The destruction of the area has been almost complete,
and these images from DigitalGlobe's Quickbird satellite
make the visual point in sobering style.

DRAMATIC INCID

PICTURE CREDITS

Jesse Allen, Earth Observatory/Laura Rocchio,
    Landsat Project Science Office: 171
Ron Beck, EROS Data Center: 170
Jaques Decloitres, MODIS Land Rapid Response
    Team, NASA/GSFC: 156, 158, 162 (t, b), 168
    (t), 193, 194, 195, 199, 200, 205, 207, 210,
    229, 235, 240, 242
Digital Globe: 248, 249, 250, 251
Earth Science and Image Analysis Laboratory,
    Johnson Space Center: 154, 184, 245, 246
German Remote Sensing Data: 152 (l)
F Hasler, M Jentoft-Nilsen, H Pierce,
    K Palaniappan and M Manyin: 182
Hubble Heritage Team (AURA, STScI, NASA): 120
ISS: 64 (r)
Landsat 7 Science Team: 163 (l), 202
Landsat 7 Science Team and NASA GSFC: 203
Courtesy Landsat 7 project and EROS Data
    Center: 179 (l)
Brian Montgomery, NASA GSFC; data courtesy
    MODIS Science Team: 201
NASA: 12, 14 (t), 14 (b), 15, 16 (t, b), 17 (l), 19
    (l), 21 (r), 23, 31 (l), 33 (r), 37 (l), 38 (bl), 38
    (br), 39, 42 (l, r), 43, 44, 45 (l), 46 (t), 47, 48
    (r), 49 (l, r), 50 (l, r), 51, 53 (l), 54 (l, r), 55,
    56, 57, 58 (l, r), 60, 61 (r), 63 (r), 64 (l, r),
    65, 66 (l, r), 67 (l, r), 68 (l, r), 69, 70 (l, r), 71,
    72, 73 (l, r), 74, 75, 76, 82 (r), 83 (l), 86 (b),
    89 (l), 100, 101, 102, 103 (l, r), 109 (l), 112,
    115, 116, 117, 122, 128, 130, 148, 151, 166,
    181 (r), 190, 192, 218, 219, 223, 233, 247
NASA/Buzz Aldrin: 38 (tl)
NASA/Aqua: 214, 215, 216, 217, 220
NASA/Neil A Armstrong: 18 (l), 22 (l, r)
NASA/Alan Bean: 25 (l, r)
NASA, Wolfgang Brander JPL-IPAC,
    Eva K Grebel: 141
NASA, Thomas M Brown, Charles W Bowers,
    Randy A Kimble: 139
NASA, A Caulet St-ECF, ESA: 140
NASA/Eugene Cernan: 31 (r), 34 (l, r), 35
NASA/Michael Collins: 19 (r)
NASA/Charles Conrad Jr: 18 (r), 20, 21 (l)
NASA, CRO: 136
NASA/Defense Meteorological Satellite
    Program: 189
NASA/Charles M Duke Jr: 27 (l) 32, 33 (bl), 36
NASA, ESA, S Beckwith (STScI) and the HUDF
    Team: 143
NASA, ESA and the Hubble Heritage Team
    (STScI/AURA): 142 (t)
NASA, ESA, HEIC and the Hubble Heritage Team
    (STScI/AURA): 142 (b)
NASA, ESA, Y Naze (University of Leige, Belgium)
    and Y H Chu (University of Illinois, Urbana): 145
NASA, ESA and A Nota (STScI): 144
NASA, Don Figer, STScI: 137 (l)
NASA, A Fruchter and the ERO Team, STScI,
    ST-ECF: 129
NASA/GSFC: 204

The NASA GSFC Scientific Visualisation Studio: 188
NASA/GSFC/LaRC/JPL, MISR team: 168 (b), 187,
    208, 209, 211, 221, 222, 226, 234, 236,
    238, 241
NASA/GSFC/MITI/ERSDAC/JAROS, and US/Japan
    ASTER Science Team: 157, 161 (r), 169, 173,
    174, 178, 206, 244
NASA JP Harrington and KJ Borkowski University
    of Maryland: 138
NASA, JJ Hester Arizona State University: 131
NASA, JJ Hester and P Scowen, Arizona State
    University: 134
NASA/The Hubble Heritage Team
    (AURA/STScI): 126, 127, 132, 133, 135, 137 (r)
NASA and the Hubble Heritage Team (STScI): 124
NASA/Hui Yang University of Illinois: 125
NASA/James B Irwin: 37 (br), 37 (tr)
NASA/ISS: 185
NASA/JPL: 48 (l), 79, 80 (t), 83, 90–91 (t), 90
    (b), 93, 106, 107, 159 (l, r), 163 (r), 165, 167
    (l), 172, 177, 179 (r), 180, 181 (l), 183.
NASA/JPL/Cornell: 78, 81, 86 (t), 87, 88,
    89 (r), 91 (b), 96, 98, 110, 111, 113, 114
NASA/JPL/Cornell/USGS: 80 (b)
NASA/JPL/GSFC/Ames: 104
NASA/JPL/Malin Space Systems: 92
NASA/JPL/NIMA: 150, 152 (r), 153, 167 (r), 186
NASA/JPL/Northwestern University: 119 (r)
NASA/JPL Ocean Surface Topography Team: 213
NASA/JPL/Space Science Institute: 84, 108,
    109 (r)

NASA/JPL/University of Arizona: 85
NASA/JPL/University of Colorado: 105
NASA/James McDivitt: 45 (r), 46 (b), 52
NASA/Edgar Mitchell: 17 (r), 33 (tl)
NASA/Harrison H Schmitt: 24 (l, r)
NASA/Russell L Schweickart: 40
NASA/David R Scott: 28 (l, r), 29, 30, 62, 63 (l)
NASA/Andy Thomas: 61 (l)
NASA/Unknown: 59
NASA, Donald Walter (South Carolina State
    University), Paul Scowen and Brian Moore
    (Arizona State University): 123
NASA/John W Young: 26, 27 (r), 53 (r)
ORBIMAGE, Inc and NASA Goddard Space Flight
    Center: 155, 160, 198, 224/5, 228,
    232, 237
Jeff Schmaltz, MODIS Rapid Response Team,
    NASA/GSFC: 196, 197, 227, 243
SOHO/Extreme Ultraviolet Imaging Telescope
    (EIT) Consortium: 99, 118
Space Station Alpha, provided by the Earth
    Sciences and Image Analysis Laboratory at
    Johnson Space Center: 164
Konstantinos Stefanidis, EO-1 Team: 176
Barbara Summey, NASA Goddard Visualisation
    Analysis Lab, based on data processed by
    Takmeng Wong, CERES Science Team, NASA
    Langley Research Centre: 212
USAF: 16 (t)
USGS Landsat 7 Team at the Eros Data
    Center: 230

ACKNOWLEDGMENTS

Many thanks to all of those who have helped in the production of this book, including Gary Kitmacher, Steve Nesbitt, Jody Russell and Steve Garber from NASA, astronaut Joseph R 'Joe' Tanner, and Neil Baber and the team from David & Charles. Thanks also to Sarah, Emilia and Charlie for all their support during the project.

WEBSITES

The many and varied NASA websites offer an unrivalled resource for educationalists, researchers and those who are simply interested in discovering what is happening at the cutting edge of space research. Many sites have facilities for high-resolution imagery and research material to be downloaded free of charge, and others offer up-to-the-minute reports on the latest missions to allow schools and individuals to keep track of progress as the exploration of space continues apace. There are so many sites available that it is impossible to list them all here, but this is a selection of some of the best, and it is up to the individual to follow the links that are offered and to explore the variety of sites on offer for themselves.

- http://spaceflight.nasa.gov (images and information from Human Space Flight)
- http://eol.jsc.nasa.gov (images of Earth taken from manned spacecraft and the International Space Station)
- http://hubblesite.org/newscenter/ newsdesk/archive/ (Hubble Space Telescope images)
- http://images.jsc.nasa.gov (collection of 9000 NASA press release photos)
- http://grin.hq.nasa.gov/subject-science. html (Great Images in NASA)
- http://visibleearth.nasa.gov/ (Visible Earth website, pictures of Earth from space)
- http://marsrovers.nasa.gov/home/index. html (Mars Exploration Rover Mission)
- http://aqua.nasa.gov/ (home page of the Aqua Satellite)
- http://terra.nasa.gov/ (home page of the TERRA satellite)
- http://www.nasa.gov/mission_pages/ cassini/main/index.html (home page of the Cassini Mission)
- http://spaceflight.nasa.gov/shuttle/ index.html (Space Shuttle home page)
- http://www.nasa.gov/mission_pages/ station/main/index.html (International Space Station home page)
- http://sohowww.nascom.nasa.gov/ (home page of the SOHO project that studies the sun)